AF906198

Remedies of Malnutritional Cancer by Edible Plants

The Author

Dr. Govind Pandey, "Professor/Principal Scientist" of Pharmacology & Toxicology, with experience of about 33 years in 'Research/Teaching/Extension/Administration', is probably "only Person in Madhya Pradesh and alone Veterinarian in India with Maximum Academic Qualifications" (20 Degrees/Diplomas/Certificates). He obtained PhD Hon. degree in Veterinary Pharmacology & Toxicology from Jawaharlal Nehru Krishi Vishwa Vidyalaya (JNKVV), Jabalpur in 1990. Presently, he is doing DSc. His "Biography" is included in the famous book of the world, "Who's Who in the World 2011" (28th edition, America). Dr. Pandey has been honoured with 3 prestigious national and international 'Fellowship Titles', viz., "FASAW, FSLSc and FISCA".

Dr. Pandey started his career as "Veterinary Assistant Surgeon/Lecturer" on 7th August, 1980 at Artificial Insemenation Training Institute, Mandla, Animal Husbandry Department, Govt. of MP. There, he also worked as "Veterinary Surgeon" and "Senior Veterinary Surgeon" in different offices at Jabalpur, including "Officer-In-Charge cum Drawing Disbursing Officer (DDO)" of Rinder Pest, Jabalpur Division, Jabalpur till 19th April, 2012. During these periods, he also served as "Chief Executive Officer/Block Development Officer cum DDO" of some Janapad Panchayats under the Panchayat & Rural Development Department, Govt. of MP; and as "Assistant Professor & Head" and "Professor/Principal Scientist & Head" of Pharmacology in Pharmacy colleges. On 20th April, 2012, he joined as "Deputy Director of Research/Associate Professor/Senior Scientist" at the Directorate of Research Services, Nanaji Deshmukh Veterinary Science University (NDVSU), Jabalpur. On 26th November, 2012, he has resumed the post of "Professor/Principal Scientist & Sectional Head", Department of Pharmacology & Toxicology, College of Veterinary Science & AH, Rewa (NDVSU, Jabalpur).

He has investigated some "Antihepatotoxic and Anticancer Herbal Drugs", and produced experimental "Hepatotoxic and Cancer Models" in animals. He has published more than 225 scientific papers. He has supervised/guided/co-guided many "Research Theses/Dissertation/Projects Reports". His 2 scientific "Books/Manuals" have been published and 6 are under publication. Dr. Pandey is the recipient 30 "Awards/ Honours/ Fellowships" (including "ICAR Senior Research Fellowship" and "Sri Ram Lal Agrawal National Award") in science, research and Hindi literature. In Hindi literature, he has published xperience.azines, A.I.R., T.V.,5 books, released 2 audiocassettes of own lyrics and edited 1 book. His several poems, lyrics, dramas or stories have been published/broadcasted through various media. He is the "Life Member" of 25 scientific, professional, literary and cultural associations/societies/journals. He has acted as the "Chairperson/Chief Guest/Judge/ Expert" in many conferences/projects/committees/programmes. He also acted as "Editor/ Mentor/Editorial Board Member/Reviewer" of many books/journals or magazines.

Remedies of Malnutritional Cancer by Edible Plants

— Author —

Dr. Govind Pandey

BVSc and AH, MVSc and AH, PhD Hon. (Pharmacol.), DSc (std.), LLB, LLM, MBA, MA (Soc.), MA (Hin.), MA (Eng.), MA (Pol.), Acharya (Jyotish), PGDPA and LSG, PGDCA, AvR, MDEH, SR, DNHE, AIT, PGPHT, FSLSc, FASAW, FISCA

Professor/Principal Scientist of Pharmacology and Toxicology, The Nanaji Deshmukh Veterinary Science University, Jabalpur, M.P., India

2014

Daya Publishing House®

A Division of

Astral International Pvt. Ltd.

New Delhi – 110 002

© 2014 AUTHOR
ISBN 9789351301653

Publisher's note:
Every possible effort has been made to ensure that the information contained in this book is accurate at the time of going to press, and the publisher and author cannot accept responsibility for any errors or omissions, however caused. No responsibility for loss or damage occasioned to any person acting, or refraining from action, as a result of the material in this publication can be accepted by the editor, the publisher or the author. The Publisher is not associated with any product or vendor mentioned in the book. The contents of this work are intended to further general scientific research, understanding and discussion only. Readers should consult with a specialist where appropriate.
Every effort has been made to trace the owners of copyright material used in this book, if any. The author and the publisher will be grateful for any omission brought to their notice for acknowledgement in the future editions of the book.
All Rights reserved under International Copyright Conventions. No part of this publication may be reproduced, stored in a retrieval system, or transmitted in any form or by any means, electronic, mechanical, photocopying, recording or otherwise without the prior written consent of the publisher and the copyright owner.

Published by : **Daya Publishing House®**
A Division of
Astral International Pvt. Ltd.
– ISO 9001:2008 Certified Company –
4760-61/23, Ansari Road, Darya Ganj
New Delhi-110 002
Ph. 011-43549197, 23278134
E-mail: info@astralint.com
Website: www.astralint.com

Laser Typesetting : **Classic Computer Services**, Delhi - 110 035

Printed at : **Replika Press Pvt. Ltd.**

PRINTED IN INDIA

Dr. Govind Prasad Mishra
MVSc (Animal Nutrition), PhD
Vice Chancellor

The Nanaji Deshmukh Veterinary Science University (NDVSU),
Jabalpur – 48 2001, M.P., India
Ph: 0761-2678007;
E-mail: vcmppcvv@gmail.com

Foreword

It is a matter of great pleasure to know that Dr. Govind Pandey of our university has written a book entitled *"Remedies of Malnutritional Cancer by Edible Plants"*. In fact, nutrition is closely associated with the health. The food eaten must not only be nutritious but it must also be balanced and clean. If the right food is not consumed as per individual requirements, it results in 'malnutrition' even to cancer. Plants serve as food and medicinal source which maintain the health and vitality of the individuals. Edible plants/foods have original place in the treatment of diseases with no ill effect. This reference book includes all such related topics.

I hope this book will provide a platform to the entire professionals engaged in the teaching, research especially in the field of Veterinary/ Animal Husbandry, Medical, Ayurvedic and Pharmacy Sciences for discussion and debate relating to the challenges and problems laced by the concerned people or animals in responding positive aspect of malnutritional cancer.

I congratulate and extend my warm wishes to author Dr. Govind Pandey for his sincere efforts in focusing the attention to such challenging issues. I am sure the book will prove to a useful publication to society.

Dr. Govind Prasad Mishra

Preface

"Remedies of Malnutritional Cancer by Edible Plants" is useful to the students, teachers and researchers of Veterinary, Medical, Ayurvedic, Pharmacy and other Life Science disciplines. In the so called the 'Third World', where poverty is common, millions of poor people are engaged in hard labours. Millions and millions of animals like horses, donkeys, bullocks and he-buffaloes, etc. work day in day out in cart or for other carriage. All of them don't get good food and don't get enough to drink. They have to work in the burning sun, on busy and dusty roads, without one moment of rest. Next to that some individuals don't even know that 'What is pain'. A lot of poor people and different species of animals have to work till they drop dead, and that often happens in the streets. They are day-to-day exposed to numerous environmental hazards and disasters, including xenobiotics, contaminants, radiations, genetic factors, etc. All these result in 'Malnutrition' which may cause 'Cancer', *i.e.*, "Malnutritional Cancer". Overall, the poverty and decreased dietary intake of good quality are the most important causes of malnutritional cancer. Now, the cancer is a burning problem in this westernized 'Ultra-Modern Era' even in our country. From ancient times we, in "Bharatvarsha" have endowed food with magical qualities. In the 'Vedic Kaal', the food was associated with divine attributes. Later, these patterns became well established. The episodes in the history of man's battle against disease clearly illustrate the

fact that food can cure many diseases. Thus, 'Foods or Edible Plants' can be the best remedies for malnutritional cancer. The medicinal plants, including fruits and vegetables have versatile immuno-modulatory and anticancer activities. Consumption of large amounts of fruits and vegetables can prevent the development of cancer. Many doctors recommend that people wishing to reduce risk of cancer eat several pieces of fruits and several portions of vegetables every day.

This work has sharpen my skills and shaped the attitudes; it encouraged me to think, to discover solutions of problems; developed interpersonal and communication skills; and gave self confidence and a sense of achievement. I would like to articulate my heartfelt gratitude to Hon'able Dr. Govind Prasad Mishra Ji, Vice-Chancellor, The Nanaji Deshmukh Veterinary Science University (NDVSU), Jabalpur, MP for expression of pleasure in good fortune towards me. I also take the opportunity to express best regards and heartiest thanks to Dr. A.B. Shrivastav, Director, Centre for Wildlife Forensic and Health, NDVSU, Jabalpur and Dr. S.P. Pandey, Professor and Head of Pharmacology, Govt. NSCB Medical College, Jabalpur for their concrete suggestions, ever-available help and encouragement. I also pay my appreciation and thanks to Dr. Madhuri Sharma for being the co-author of many published papers related to this book.

Lastly, "All beings are fond of themselves, they like pleasure, they hate pain, they shun destruction, they like life and want to live long. To all, life is dear; hence their lives should be protected".

Dr. Govind Pandey

Contents

1
Introduction

'Neoplasm or tumour' is a mass of tissues formed as a result of abnormal, excessive, autonomous and purposeless proliferation of cells. "Cancer" (malignant tumour), a disease of human and other multicellular animals may be uncontrollable and incurable, and may occur at any time at any age in any part of the body. In other words, cancer is a frightful disease because the patient suffers pain, disfigurement and loss of many physiological processes. It is caused by a complex, poorly understood interplay of genetic and environmental factors[1–6]. Cancer continues to represent the largest (or second leading) cause of mortality and claims over 6 millions people yearly in the world. On the other hand, the incidence of cancer is increasing worldwide and annually about 3500 per million human population are died globally[2–5,7–9].

In fact, cancer may be caused in one of the three ways: incorrect diet, genetic predisposition and via the environment. At least 35 per cent of all cancers worldwide are caused by incorrect diet, and in the case of colon cancer, diet may account for 80 per cent of the cases. When one adds alcohol and cigarettes to the diet, the percentage may increase to 60 per cent. The genetic predisposition gives rise to 20 per cent cancer cases. Thus the majority of cancers are being associated with a host of environmental carcinogens[3–4,8,10–13]. As long ago as 480 BC, Hippocrates recognized that several aspects of what we now call 'lifestyle' must come together to produce a healthy body. He said, 'Positive health requires a knowledge of man's primary constitution and the powers of various foods, both those natural to them and those resulting from human skill'. What Hippocrates called "man's primary constitution", we today call 'genetics', and we can infer that foods 'resulting from human skills' can be equated with today's diet[3,8,11–13].

"Nutrition" is a scientific discipline which essentially deals with the food and related issues. The simple meaning of nutrition is the study of what happens to food once it is being consumed by the individual. Nutrition is closely associated with health. If there is any imbalance in eating pattern, *i.e.*, eating too little or too much will result in poor health. The food eaten must not only be nutritious but it must also be complete and clean, otherwise the individual eating the food would get ill even the food is nutritious. So, if the right food is not consumed in right amounts, it results in "malnutrition" (mal-noo-trish-un) which may result in either 'undernutrition' or 'overnutrition'. One example of undernutrition is vitamin A deficiency when young children do not consume enough vitamin A rich foods, hence blindness results. An example of overnutrition is obesity, which occurs due to excess consumption of energy leading to accumulation of fat in the body[3–4,8,12,14].

Malnutrition occurs when we do not get enough calories or nutrients and hence, it may cause several diseases, including cancer. Malnutrition acts as carcinogen in many ways to produce cancer in human beings and animals, and so the aetiological role of malnutrition preceding clinical cancer has been firmly established[3–4,8,15]. Malnutrition is directly responsible for 300,000 deaths per year in children younger than 5 years in developing countries and contributes indirectly to over half the deaths in childhood worldwide. Globally, 113.4 million children younger than 5 years are affected as measured by low weight for age. The overwhelming majority of these children, 112.8 million will live in developing countries with 70 per cent of these children in Asia, particularly the south-central region and 26 per cent in Africa. An additional 165 million (29 per cent) children will have stunted length/height secondary to poor nutrition[3–4,8,16]. Thus, malnutrition is the single largest cause to the high rate of infant and child mortality and morbidity in India. The single major factor responsible for the wide prevalence of malnutrition in India is the poverty. Poverty and lack of food are the primary reasons why malnutrition occurs in the United States. Ten per cent of all members of low income households do not always have enough healthful food to eat, and malnutrition affects one in four elderly Americans. About 50 per cent of our people live below the poverty line and even after spending 80 per cent of their income on food, they cannot have a balanced diet[3–4,8]. Studies show that child malnutrition is closely linked to income level, the widely used poverty indicator. That the relationship between child malnutrition and poverty is most sensitive at the lower end of the income range makes child malnutrition a good indicator for development intervention projects and programmes[3]. The situation is similar in animals.

Like many countries, India is also facing a great problem of malnutrition. Thus the children suffering from malnutrition need special attention. Those in moderate and mild degree of malnutrition require additional food in order to bring their weight to normal; whereas, those children suffering from a severe degree of malnutrition need special attention by a health functionary as they usually suffer from diarrhoea, respiratory infections and other diseases. Malnutrition predisposes a person to infection. On the other hand, infections lead to malnutrition. Human body requires energy to carry out the different activities, and proteins are necessary for growth and repair. Deficiency of these two basic nutrients in our body leads to "protein energy malnutrition" (PEM). This disorder has very serious consequences, leading to cancer and/or death. PEM is the major nutritional problem in India and is widely prevalent among children (0-6 years) but is also observed as starvation in adolescents and adults, mostly lactating women, especially during periods of famine or other emergencies[3–4,8,17–18].

"Anaemia" is the reduction below normal of circulating red blood cell (RBC). Anaemia (anemia) is not a disease in itself but a symptom. It may be caused by blood loss, red blood cell destruction (known as haemolysis) or inadequate RBC production. Blood loss and RBC destruction are classified as "regenerative anaemia". Inadequate RBC production is classified as "non-regenerative anaemia". This is where the body is unable to produce more RBCs in the bone marrow. Both malnutrition and anaemia in humans (particularly in young children) and animals are substantial problems in India. Both these are very closely related with each other, and they in turn may lead to the development of cancer in due course of time. The animals, particularly horses, cattle, goats, dogs and cats also suffer from malnutritional anaemia ultimately resulting into various types of cancer.

It is well known fact that poor diets can cause diseases or on the other hand, some diseases can be cured or controlled by using particular types of diets. From ancient times we, in India, have endowed food with magical qualities. In the Vedic times, food was associated with divine attributes. This tradition was prevalent in other societies as well- ancient Egypt, for example. Ancient people discovered the healing powers of food- perhaps initially by accident. Later, these patterns became well established and were described in medical texts. The astonishing fact is their proven value in treating disorders primarily caused by lack of food and nutrients. Another dramatic episode in man's search for a cure is the story of scurvy, which is caused by vitamin C deficiency. Fresh herbs, oranges and other citrus fruits were found to be effective in preventing scurvy. These episodes in the history of man's battle against disease clearly illustrate that food can cure certain diseases. It is quite evident, however, that the diseases which can be cured by food or nutrient concentrates are caused by deficiency of particular nutrients. Disorders can be cured by giving food sources rich in the nutrient that is otherwise lacking in the diet. Thus there is no doubt that food can help to control several ailments[3,8,19–20].

Treatment of malnutrition depends on its causes and severity. The treatment may include increasing the calories and nutrients in diet. Although eating disorders (malnutrition) are treatable and many people recovered, but the recovery is a complex process that can take several months or even years. If malnutrition results in cancer, it is very difficult to treat, but some improvements can be brought about by taking proper foods or nutrients and drugs[3,8,20–21]. Modern (allopathic) drugs being a major treatment modality for the treatment of diseases, exhibit severe toxicity on normal tissues. Thus there is a definite need to search alternate drugs to be effective in the prevention and cure of diseases caused by malnutrition. Because of variations in the aetiology of cytotoxicity and cancer, there is a great difficulty to diagnose them, and a physician is rarely able to provide any specific treatment. At the most, supportive and symptomatic treatments are given but the multiplicity of deranged functions renders the treatment still more complicated[3,8–9,20,22–23].

The plant kingdom plays a major role in the life of human beings and animals. It serves as a food and medicinal sources, which maintain the health and vitality of the individuals. Edible plants/foods have original place in the treatment of diseases with no ill effect. India is the largest producer of medicinal plants and is rightly called the "Botanical garden of the World". In the indigenous or traditional system of medicine, numerous medicinal plants and their formulations are now being used for the treatment and control of various diseases, including cancer[2–3,5–6,8–9,11–12,20,22–24–25]. More than 50 per cent of all modern drugs in clinical use are of natural products, many of which have the ability to include apoptosis in various tumour cells of human origin[5,8,26]. According to WHO estimates, more than 80 per cent people in developing countries depend on traditional medicine (indigenous or herbal drugs) for their primary health needs. A recent survey shows that more than 60 per cent of patients use vitamins or herbs in their therapy[5,8–9,11,27–29].

Medicinal plants are known to have good immunomodulatory properties. They act by stimulating both non-specific and specific immunity. They may promote host resistance against infection by re-stabilizing body equilibrium and conditioning the body tissues[2,5–6,8–9,12,25,30–31]. Some medicinal plants and their products, including vegetables, fruits and crops play an important role for cancer prevention. Consumption of large amount of vegetables and fruits may prevent from the cancer. Many doctors recommend that people wishing to reduce their risk of cancer eat several pieces of fruits and several portions of vegetables every day. Many plant-derived products exhibit potent anticancer activity against several cancer cell lines Thus, the medicinal plants, including edible plants (fruits and vegetables) elicit immunomodulatory and anticancer activities[8–9,20,27–28,32].

2
Malnutrition and Cancer

Malnutrition

If an individual eats the right kind of foods in the required amounts, that individual will keep good health provided no other factors intervene. On the other hand, a poor eating pattern or eating too little or too much will result in poor health. These are both facets of "malnutrition". When the diet supplies too little of one or more nutrients we suffer from a particular form of malnutrition called 'undernutrition'. When the diet provides too much of one or more nutrients, another form of malnutrition called 'overnutrition' results. Nutritional disorders like obesity, heart disease and diabetes mellitus are the results of overnutrition[3,4,8,19]. Therefore, malnutrition is a general term for the medical condition caused by an improper or insufficient diet. It most often refers to undernutrition resulting from an inadequate consumption, poor absorption, or excessive loss of nutrients, but the term can also encompass overnutrition resulting from overeating or excessive intake of specific nutrients. An individual will experience malnutrition if the appropriate amount of, or quality of nutrients comprising a healthy diet are not consumed for an extended period of time. An extended period of malnutrition can result in starvation or many other diseases such as scurvy and cancer[3,8,33].

Malnutrition is not a new health problem in India. As far back as 1946, when the Bhore Committee reviewed the health status of the Indian population, diet surveys had already been done. These surveys showed that about 30 per cent of the families surveyed consumed inadequate amounts of food to provide the necessary nutritional requirements. At that time, it was also known that an adequate diet contributed to high mortality and morbidity in the general population particularly in infants and pregnant women[3–4,8,34]. Children suffering from malnutrition almost always have infections such as diarrhoea and respiratory infections. Just as infections are common in cases of severe malnutrition, infections like measles, diarrhoea and whooping cough can also cause malnutrition. Thus, malnutrition can increase the risk of infections and infections can, in turn, lead to malnutrition. This inter-relationship and synergistic effect of malnutrition and infections often lead to a high incidence of cancer and/or deaths. Malnutrition leads to infection by reducing the diseases-fighting capacity of the body. Infection on the other hand leads to malnutrition due to loss of appetite, dietary restriction, reduced absorption, etc.[3–4,8,17]. There are many environmental or pollution factors, including poverty that contribute to the development of an eating disorder or malnutrition (Figures 1–6[35]).

The people who are at greater risk of having malnutrition include:

1. Infants and children who do not get the right type or amount of calories and nutrients;
2. Pregnant women;
3. Older adults;
4. People who are taking medicines that decrease appetite or affect the digestion and absorption of nutrients;
5. People who abuse drugs or alcohol;
6. People with eating disorders;
7. People with certain diseases such as cystic fibrosis, celiac disease, liver disease, kidney disease and cancer; and
8. People with a low income who have trouble to buy the right kind of food.

It has already been mentioned under the "Introduction" that due to deficiency of energy and protein, PEM results[3–4,8,17–18,36]. Hence, PEM can be defined as a range of pathological conditions arising from a deficiency

Figure 1

Figure 2

Figure 3

Figure 4

Figure 5

Figure 6

Figures 1–6: Environmental factors: Hazardous child labour pollution–Poverty income generation malnutrition

[*Source*: Stock photos of malnutrition.htm, 2011]

of protein and energy, and is commonly associated with infections. On the other hand, PEM is a disease of the poor, undernourished and chronically ill patient characterized by imbalance between the supply of nutrient and energy and the body's demand for them to ensure growth, maintenance of health and function of the body. PEM mostly occurs in children below 5 years and failure to thrive is a common presentation. This disorder is of three types, namely:

(*a*) Marasmus- Deficiency of both energy (carbohydrate) and protein;

(*b*) Kwashiorkor- Deficiency of only protein;

(*c*) An intermediate state of marasmus and kwashiorkor.

"Marasmus" is a Greek term, meaning 'to waste'. It is characterized by very low body weight and weight loss due to growth failure, stunting (short in height for age), loss of subcutaneous fat, gross muscle wasting, emaciation and looking like a baby monkey, and failure to thrive. In addition, watery diarrhoea associated often with dehydration, irritation (does'nt allow to touch), apathy, anxiety, decreased responsiveness, behavioural changes, no oedema, wrinkled skin dry and loose (like tissue paper), and loss of even buccal pad of fat are also prominent features. The person may have other deficiencies particularly, vitamin A deficiency. Marasmus occurs more frequently in infants and very young children (Figures 7[36] and 8[35]).

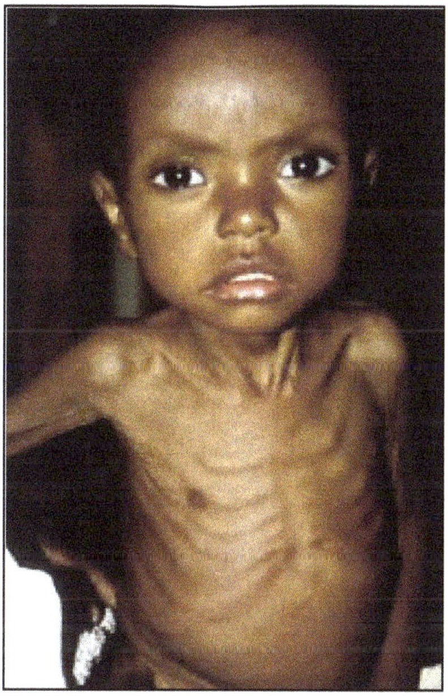

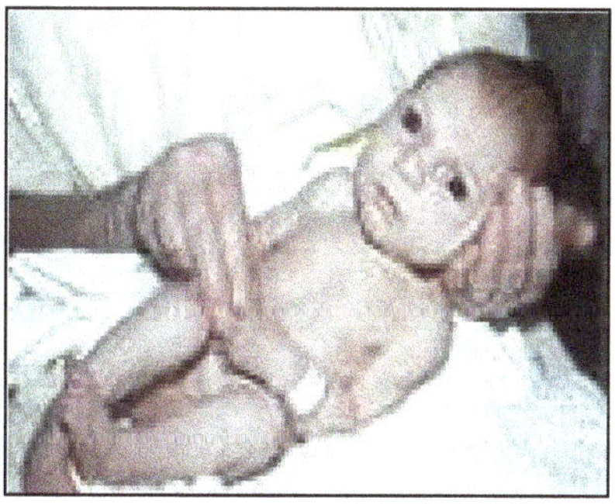

Figures 7 and 8: Marasmus PEM
[*Source*: Protein Energy Malnutrition.htm, 2011].

"Kwashiorkor" is also called the 'disease of weaning' as it appears during the period of weaning in a child when the mother fails to supplement the proteins required but feeds the cereals and malt which are rich in carbohydrate but poor in proteins. It is characterized by oedema of whole body especially belly, wasted muscles, and very low body weight for age due to growth failure (retarded growth). Other signs and symptoms include psychomotor changes, hepatomegaly (enlarged, fatty liver), irritability, moon face, hair changes (flag signs), skin changes (flaky paint appearance and mosaic skin appearance), thin, soft and fissured nail plates, vitamins A and B deficiencies, and associated diseases such as watery diarrhoea, severe respiratory infection (cough), etc. Kwashiorkor syndrome is most frequently observed in children aged 1 to 3 years and is precipitated by an infection or more commonly by a series of infections (Figures 9[36] and 10[35]).

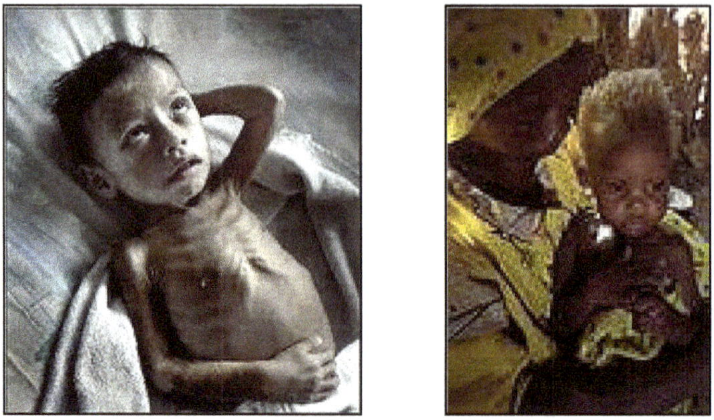

Figures 9 and 10: Kwashiorkor PEM
[*Source*: Stock photos of malnutrition.htm, 2011].

In intermediate type of PEM, the features of both marasmus and kwashiorkor PEM are seen. Associated symptoms like vitamin C deficiency, zinc deficiency, atrophy of the papillae on the tongue, angular stomatitis, xerophthalmia and cheilosis may occur[36].

PEM can result in marked hypoalbuminemia, anemia, oedema, muscle atrophy, delayed wound healing, impaired immunocompetence, weight loss, reduced basal metabolism, depletion of subcutaneous fat and tissue, bradycardia and hypothermia. The patient with PEM is typically emaciated, elderly and chronically ill. Thus malnutrition places stressed patients at a greatly increased risk for morbidity, leading to cancer and mortality. Some complications are more common in malnourished patients and those with inadequate nutrient intake than in well-nourished individuals. These complications may be a longer recovery period, impaired host defenses and sepsis, impaired wound healing, anemia, impaired gastrointestinal tract function, muscle atrophy, impaired cardiac function, impaired respiratory function, reduced renal function, brain dysfunction, delayed bone callus formation and atrophic skin[3,18].

Malnutritional Cancer

Cancer (malignant neoplasm) is a class of disease in which a group of cells display uncontrolled growth through division beyond normal limits, invasion that intrudes upon and destroys adjacent tissues, and sometimes metastasis, in which cancer cells spread to other locations in the body via lymph or blood. These three malignant properties of cancers differentiate them from benign tumours, which are self-limited, and do not invade or metastasize. Cancer is primarily an environmental disease with 90-95 per cent of cases due environmental factors such as lifestyle, and 5-10 per cent directly due to heredity. Common environmental factors leading to cancer include: tobacco (25-30 per cent), diet and obesity (30-35 per cent), infections (15-20 per cent), radiation, lack of physical activity, and environmental pollutants. These environmental factors cause or enhance abnormalities in the genetic material of cells. Cell reproduction is an extremely complex process, which is normally tightly regulated by several classes of genes, including oncogenes and tumuor suppressor genes. Hereditary or acquired abnormalities in these regulatory genes can lead to developmemt of cancer[36].

Abnormal diets such as fad or limited diet and recent decrease in food intake usually cause malnutrition[18], which may lead to the development of cancer. There is an increased risk of malnutrition associated with chronic diseases such as cancer, and diseases of intestinal tract, kidney and liver. Patients with these chronic diseases, especially cancer may lose weight rapidly and become susceptible to undernourishment because they can not absorb valuable vitamins, calories and iron. Studies show that child malnutrition is closely linked to income level, the widely used poverty indicator. The relationship between child malnutrition and poverty is most sensitive at the lower end of the income range[3,8] (Figures 1–6).

People with drug or alcohol dependencies are also at increased risk of malnutritional cancer. These people tend to maintain inadequate diets for long periods and their ability to absorb nutrients is impaired by the alcohol or drug's effect on body tissues, particularly the liver, pancreas and brain. Unintentionally losing 10 lb (4.5 kg) or more weight may be a sign of malnutrition. People who are malnourished (Figures 1–11) may be skinny or bloated; their skin is pale, thick, dry, and bruises easily; rashes and changes in pigmentation are common; hair is thin, tightly curled and pulls out easily; and joints ache and bones are soft and tender. The gums bleed; and tongue may be swollen or shriveled and cracked. Visual disturbances are night blindness, and increased sensitivity to light and glare. Other symptoms include: anaemia, diarrhoea, disorientation, goiter (enlarged thyroid gland), loss of reflexes and lack of coordination, muscle twitches, scaling and cracking of the lips and mouth[3,8,37]. Malnutrition usually develops in stages over a long period of time. Signs and symptoms may not appear until the later stages of malnutrition. Some early signs and symptoms of malnutrition may include: irritable (bad mood) and tired, slower growth than normal, or no growth, and weight loss. Bone or joint pain, weak muscles, bloated abdomen, swelling in other parts of the body, change in the colour of skin and hair, brittle and spooned nails, dry and scaly fingures (Figure 11[35]) or skin, hair loss, sunken temples (sides of the head), loss of appetite, slow wound healing, and easily get infections are noticed. All these signs and symptoms are correlated with those of many cancers. There is an increased risk of malnutrition associated with chronic diseases, *e.g.*, diseases of intestinal tract, kidney and liver (Figure 12[35]), and cancer (Figure 13[38]). Patients with these chronic diseases, especially cancer may lose weight rapidly and become susceptible to undernourishment because they can not absorb valuable vitamins, calories and iron[3,8]. Due to hormonal malnutrition, cerebral (brain) atrophy-cellular ischemia may develop (Figure 14[35]).

In relation to cancer, malnutrition needs to be defined comprehensively as the carcinogenic effects produced by nutritional

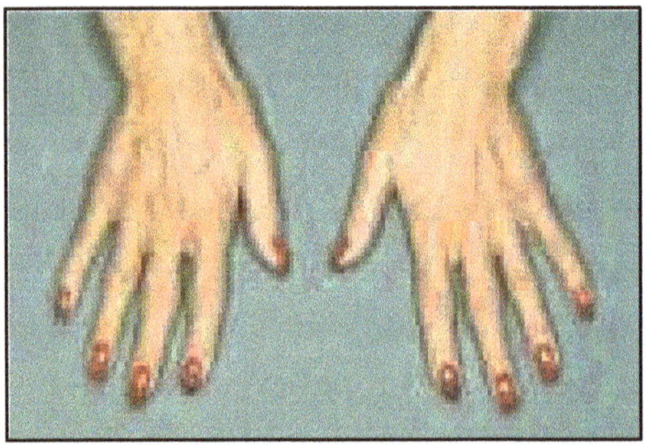

Figure 11: Woman showing spooned nails and dry scaly fingures due to malnutrition (vitamin deficiency).

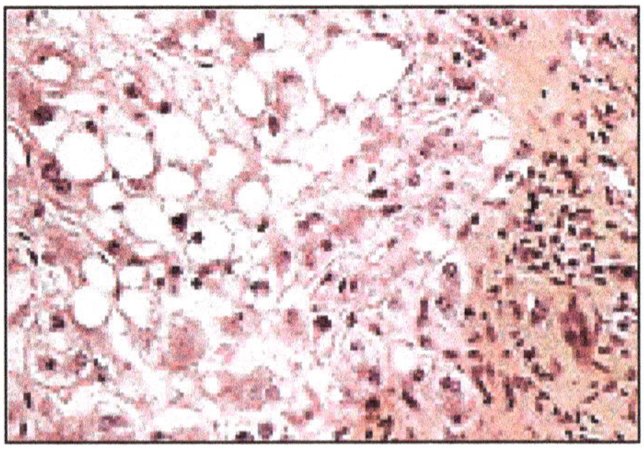

Figure 12: Liver cirrhosis due to malnutrition.

[Source: Stock photos of malnutrition.htm, 2011].

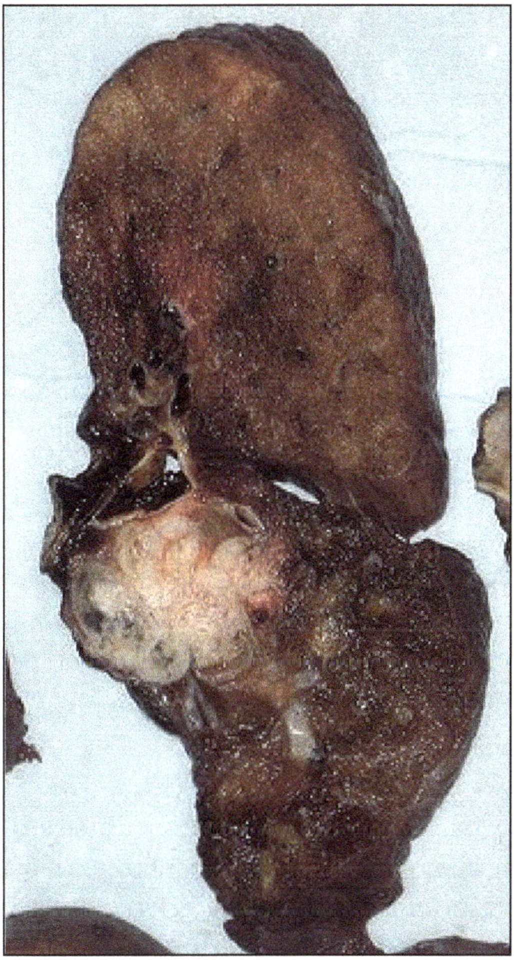

Figure 13: Cancer of .... due to hormonal malnutrition.

[*Source*: Photo of Cancer- RightHealth.htm, 2011].

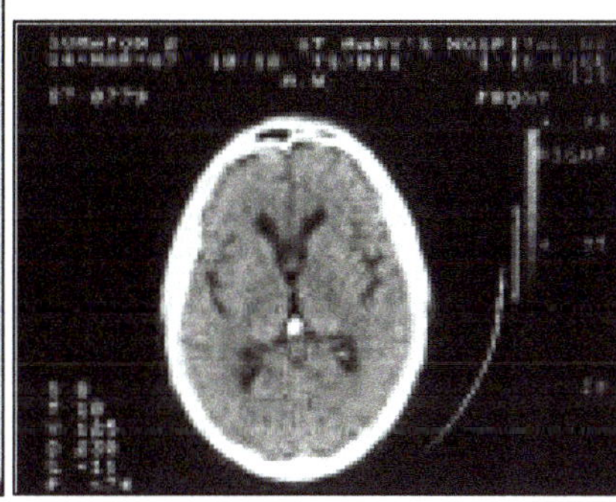

Figure 14: Cerebral (brain) atrophy-cellular ischemia due to malnutrition.

[*Source*: Stock photos of malnutrition.htm, 2011].

variables through multiple endogenous involvements. Nutritional factors act as primary effectors in four situations: carcinogens in food articles, affected bioavailability of nutrients, non-nutritive dietary items and harmful contaminants. Malnutrition acts as carcinogen in many ways to produce cancer in humans. Nutritional carcinogenesis may occur due to ingestion of toxins, malnutrition, non-bioavailability of the micronutrients and inactivation of the metabolic enzymes (*i.e.*, mixed-function oxidases) present in the liver. Thus the aetiological role of malnutrition preceding clinical cancer has been firmly established[3,8,15].

The eating habit of humans is one of the most important causes of malnutrition. The cartoon (Figures 15–16[39]) is a show that how men are not taking the meal timely and properly as required by the body.

Figures 15 and 16: Cartoon pictures on eating habits of humans. [*Source*: Malnutrition Cartoons and Comics.htm, 2011].

Decreased dietary intake is one of the most important causes of malnutrition in patients undergoing hemodialysis. Such patients often have anorexia due to reduced sense of taste and food palatability as a result of dietary sodium and water restrictions[3,40]. The risk of developing malnutrition is increased during dialysis through the loss of nutrients such as amino acids, glucose and water-soluble vitamins into the dialysate. Abnormal metabolism of nutrients, minerals and hormones also occurs in patients undergoing hemodialysis as their renal function declines. The decreased response to growth hormones and insulin-like growth factors in these patients may lead to an increase in catabolism and promote protein wasting. Also, hyperparathyroidism that occurs in patients with chronic renal failure may increase metabolism of muscle proteins and amino acids and further worsen the wasting syndrome. Metabolism of amino acids is also affected by renal insufficiency, leading to an abnormal plasma amino acid profile and impaired protein synthesis. Metabolic acidosis may increase catabolism and promote protein breakdown. This may in turn further impair nutritional status. The hemodialysis treatment is often catabolic and can promote wasting. In addition, bioincompatibility of the dialysis membrane results in cytokine production, which causes an increase in protein breakdown. Concurrent illnesses and acute or chronic conditions, *e.g.*, inflammation, infection, liver disease (Figure 12) and diabetes mellitus may compromise the nutritional status[3,41]. In chronic stage, all these disorders may lead to the development of various types of cancer (Figure 13).

Certain Environmental Substances as Carcinogens

There are certain environmental substances (contaminants or hazards) described below, which can cause either malnutrition or cancer, or malnutritional cancer as a whole both in human beings and animals[4,42].

Several reports indicate that environmental and food contaminants, including xenobiotics (environmental or foreign chemicals) such as oestrogenic endocrine disruptors (xenoestrogens), certain therapeutic agents (*e.g.*, antibiotics, antihypertensive drugs, etc.) may cause various types of cancer[43]. The 'endocrine disruptors' influence the normal functions of oestrogens, thereby cause the cancers of several organs. Chemicals, industrial wastes, pesticides, oral contraceptives (OCs) or oestrogens, detergents, food additives and plastics, all are sources of environmental toxins and endocrine disruptors. The potential endocrine disrupters such as antibiotics, hormones, plasticizers and non-ionic surfactants are becoming the priority pollutants. Most of these have been shown as carcinogenic[42–44] or they may contaminate the diet or food, thereby causing malnutrition[3].

Many factors have a major effect on increasing the rates of oral, colon, lung and breast cancers. Some of these factors include increased infections, more use of pesticides, low consumption of fruits and vegetables, increased consumption of alcohol and red meat, more smoking, high industrial pollution, more exposure to sun, decreased physical activity and high occupational exposures. The major portion of chemicals to which humans are exposed is naturally occurring, that are carcinogenic at large doses. Almost every fruit and vegetable contains natural carcinogenic pesticides. Many natural chemicals are ingested as carcinogens from cooking food like roasted coffee. They include caffeic acid, catechol, dichlorodiphenyltrichloroethane (DDT), furfural and benzo(a)pyrene, etc. Food additives (*e.g.*, allyl isothiocyanate, alcohol, butylated hydroxyanisole and saccharin), mycotoxins (*e.g.*, aflatoxin and hepatitis B virus) and synthetic contaminants (*e.g.*, polychlorinated biphenyls and tetyrachlorodibenzo-p-dioxin) are also mutagenic and carcinogenic[45]. Aryl hydrocarbons such as dioxins, polychlorinated biphenyls (PCBs) and polycyclic aromatic hydrocarbons (PAHs) are bound at the cellular aryl hydrocarbon receptor (AhR), leading to the toxicity and carcinogenicity[42–43].

Humans and animals both are exposed to dioxins (belong to a group of halogenated aromatic hydrocarbons) mainly through contaminated foods. A dioxin compound, 2,3,7,8-tetrachlorodibenzo-para-dioxin (TCDD) is the most toxic congener. The TCDD, benzo(a)pyrene and PCBs can activate AhR, which subsequently induce expression of cytochrome P4501A1 (CYP1A1) and P4501B1 (CYP1B1) enzymes. The oxidative stress caused by induction of cytochrome P450s is one of the toxic effects of dioxin. Pesticides such as insecticides (viz., carbaryl, carbosulfan, DDT, methoxychlor, dieldrin, endosulphan, sumithrin, organophosphorus compounds such as malathion and parathion, etc.) and herbicides (viz., alachlor, metolachlor, quarternary ammonium, atrazine, pentachlorophenol, etc.) have become a part of environmental contaminants due to their widespread use in agriculture and disease control. Many pesticides are immunotoxic and found to suppress the cell-mediated immunity[44]. Pesticides such as DDT and alachlor can behave like endogenous oestrogen and function to suppress apoptosis in estrogen receptor (ER)-positive human breast cancer. Some other pesticides and related chemicals may act as carcinogens. These xenobiotics adversely affected the lymphocyte function[42–43], and increased the oxidative stress and lipid peroxidation in various tissues[46].

Epidemiological studies indicate a close association between the process of westernization and an increase in breast cancer incidence. Western countries showed a higher breast cancer risk due to environmental influence. The lifestyles, including oral or dietary habits are responsible for the incidence of many cancers. Diet having several compounds, either as dietary components or contaminants or formed during food processing, has influence on the development of cancer. Mukherjee *et al.*[43] and Pandey and Madhuri[42] reported that zearalenone (a nonsteroidal compound produced by many *Fusarium* fungi species) can contaminate dairy products and cereals (*e.g.*, barley, corn, maize, rice and wheat). Zearalenone may be a potential promoter of breast tumurigenesis. After consumption, food mutagens undergo metabolic activation or detoxification by different endogenous enzymes. Most mutagens begin their adverse effects at the DNA level by forming DNA adducts with carcinogenic metabolites. The N-nitro compounds, *viz.* N-nitrosamines and nitrosamides (*e.g.*, N-nitrosureas) are a large group of chemicals that are linked with the cancer pathogenesis. Humans are exposed to N-nitrosamines from a variety of foods and tobacco smoke. Some pesticides like atrazine can be converted into genotoxic N-nitrosamines (N-nitrosoatrazine) in the environment or digestive system. PAHs are environmental hazards formed during incomplete combustion of coal, oil, gas, garbage and other organic substances such as tobacco and different food items. Leafy vegetables can be a significant source of PAHs in the human diet. In general, PAHs occur in lower amount in cigarette smoke; human exposure is predominantly from dietary sources. Metabolic activation of PAHs results in DNA binding products. Thus, the westernized dietary habits, including smoking may increase the breast cancer risk. Many lipophilic carcinogens of tobacco smoke, *e.g.*, PAHs like benzo(a)pyrene and N-nitrosamines can be stored in breast tissues. Some PAHs such as 7,12-dimethylbenz(a)athracene (DMBA) and dibenzo(a,l)pyrene are potent mammary carcinogens.

Tobacco has unfortunately become a routine part of the personal environment in the world. In India, 40 per cent of total cancer cases are tobacco related. Almost one fourth of the India's population (more than 250 million people) consumes tobacco. More than 10 million children below 15 years are addicted to tobacco. According to WHO, head, neck, lung, throat, urinary bladder, kidney, uterus and colon cancers are mostly caused by tobacco. Bidi is said to be more harmful than cigarette. Smoke tobacco contains more than 4,000 chemical compounds and at least 400 toxins, many of them are known carcinogens and irritants. Some airborne contaminants are PAHs, carbon monoxide and nitrogen dioxide, etc. Tobacco related chewing materials include pan (betal), pan masala, gutka, zarda, snuff, tobacco powder, areca nut and naswar, etc. These substances may increase the risk of oral cancer. Pan masala is unfortunately charged with polyaromatic hydrocarbons, organochlorine pesticides (DDT and

BHC), narcotics, metals and minerals. About 15 per cent of oral and pharyngeal cancers may be attributed to dietary deficiencies or malnutrition. Similarly, fungi (*Candida albicans*), viruses (human papilloma virus, hepatitis B virus, human T cell lymphotrophic virus and human immunodeficiency virus), ultraviolet (UV) radiation, immunosuppression, certain occupations (farming and industries) and poor socio-economic status aids in the development of precancerous lesions or carcinoma[42,47].

Endocrine disruptors are xenobiotic (environmental/foreign) chemicals that adversely interfere with the natural functions of hormones[42–44]. Oestrogenic endocrine disruptors or xenoestrogens are widely distributed in environment. Several chemicals such as PCB congeners, pesticides (*e.g.*, dieldrin and endosulphan), food-related mycotoxin zearalenone and its derivatives, UV screen and even some metals like cadmium can influence hormonal responses by binding to ER. The xenoestrogens interact with the binding pocket of the ER because they have chemical similarities to oestrogen (usually a phenolic A-ring). Many xenobiotic chemicals can interact with the enzyme systems that metabolize oestrogens; and by this process they may modulate the endogenous metabolism. The AhR or dioxin receptor has involvement with ER-mediated response pathways. Insecticides (*e.g.*, sumithrin) and pentachlorophenol alter steroid signaling pathways. Other man-made chemicals acting as endocrine disruptors (related with oestrogens) are atrazine, DDT, oestrogens (*e.g.*, oestradiol and ethinyl oestradiol), hexachlorophene and toxaphene, etc. All these can influence the physiological functions of oestrogen and modulate the non-genomic pathways[42–43], which may cause various malnutritonal cancers[4]. Animals are also exposed to all these substances, which can cause malnutritional cancer[20].

3
Malnutritional Anaemia Leading to Cancer

"Malnutrition" and "anaemia" in humans (particularly in young children) and animals are substantial problems in India. Both these malnutrition and anaemia (anemia) are very closely related with each other, and they in turn may lead to the development of cancer in due course of time. Hence, this chapter deals with the brief and concised knowledge of anaemia, especially resulting from the malnutrition or vice-versa, the anaemic individuals may become malnourished. In both these situations, the affected individuals can suffer from the cancer.

Anaemia: Its Types, Causes, Signs and Symptoms

Blood is composed of several different types of cells, the red blood cell (RBC) being the most common. The role of RBCs is to carry oxygen to the tissues, so the symptoms of anaemia are due to insufficient oxygen in the blood and tissues. Anaemia is, therefore, defined as a reduction below normal of circulating RBCs. Anaemia is not a disease in itself but a symptom. It may be caused by blood loss, red blood cell destruction (known as haemolysis) or inadequate RBC production. Blood loss and RBC destruction are classified as "regenerative anaemia". In regenerative anaemia, the bone marrow is able to produce more RBCs although this may not occur quickly enough to replace the lost RBCs. Inadequate RBC production is classified as "non-regenerative anaemia". This is where the body is unable to produce more RBCs in the bone marrow[48]. Anaemias can usually be broadly divided into "regenerative" or "non-regenerative" types. Regenerative anaemias are those in which the bone marrow responds to the anaemia appropriately by producing new RBCs. These can arise as a result of blood loss (haemorrhage) or haemolysis (when RBCs are broken down within the bloodstream). "Non-regenerative anaemias" are characterised by the absence of new RBC production in the bone marrow and typically reflect underlying systemic disease or bone marrow abnormalities[49].

Further, it is necessary to discuss more about anaemia. The anaemia is an abnormally low number of RBCs (also known as erythrocytes) circulating in the bloodstream. It is a condition in which the number of RBCs circulating in the blood is below the normal range for an animal species, breed and age. There are two main types of blood cell: (a) Red cell or RBC; (b) White cell or white blood cell (WBC, also known as leucocyte). The RBCs are produced by specialist cells in the bone marrow. One of the main roles for RBCs is to carry oxygen around the body. The oxygen which is breathed into the lungs gets bound to haemoglobin in the RBCs, and in the body's tissues the oxygen is released. If there are not enough RBCs in circulation, the tissues get starved of oxygen (called "hypoxia") with serious consequences, because oxygen is essential for normal cell function. There are three main causes of anaemia: (i) Individual loses RBCs due to bleeding at a rate that exceeds its ability to produce more RBCs from the bone marrow. Sometimes, bleeding is obvious but on other occasions it may not be. Bleeding externally is obvious; the bleeding into urine is usually noticed as a pink or red discolouration; and the bleeding into the large intestine (*e.g.*, colon) may be seen as fresh red blood in a stool, but the bleeding high up in the gut (in the small intestine) gets changed producing a black sticky stool called "melaena". The vomited blood often looks brown in colour like coffee; (ii) Bone marrow is unable to produce adequate numbers of RBCs to maintain adequate circulating numbers. If the disorder of bone marrow also affects WBC production, there can be low WBCs in the circulation (called "leukaemia") as well; (iii) Sufficient RBCs are being produced but they are being destroyed in the bloodstream and have a shorter lifespan. This occurs, for example, when the body produces antibodies against its own RBCs causing them to disintegrate (called "haemolysis"). This disorder is called "autoimmune haemolytic anaemia". Haemolysis is also due to the infections, including "feline infectious anaemia" caused by *Haemobartonella felis* (common in cats) and *Haemobartonella canis* (uncommon in dogs). The signs of anaemia that individuals may notice include: (1) Pale gums and other visible mucous membranes; (2) Poor ability to exercise-sometimes collapse if anaemia is severe; (3) Weakness; and (4) Increased heart rate and increased breathing rate[50].

Anaemia in Human

"Malnutrition" and "anaemia" in young children are substantial problems in India. In young children, malnutrition puts them at a higher risk of experiencing health problems such as stunted growth, mental retardation and increased susceptibility to infectious diseases. Almost half of children (48 per cent) under five years are stunted, that is too short for their age, an indicator or chronic malnutrition, and 43 per cent are underweight. The proportion of children who are severely undernourished is also notable- 24 per cent people are severely stunted and 16 per cent are severely underweight. Wasting, defined as an abnormally low weight for the child's height, is also a serious problem in India, affecting 20 per cent of children under five years of age. Very few children under five years are overweight. Overall, girls and boys are about equally likely to be undernourished. Undernutrition is substantially higher in rural areas. Even in urban areas, however, 40 per cent of children are stunted and 33 per cent are underweight. Most children aged 6 to 59 months in India (70 per cent) have anaemia. In fact, that in all but four states (Goa, Manipur, Mizoram and Kerala) more than half of children are anaemic. In India, anaemia is primarily due to poor nutrition. Anaemia is widespread in every state. Over half of women and 24 per cent of men suffer from anaemia, and have lower than normal levels of blood haemoglobin. Anaemia has increased in ever-married women from 52 per cent to 56 per cent. Among pregnant women, anaemia has increased from 50 per cent to 59 per cent. While anaemia is common throughout India, the prevalence is highest in all states in the East Region, especially Jharkhand and Bihar where more than two-thirds of women and one-third of men are anaemic. Severe anaemia is most prevalent in Assam for both women and men. Almost half of Indian women and 44 per cent of men are either too thin or too fatty, which indicates a dual burden of malnutrition. Malnutrition and anaemia are common among Indian adults. Both malnutrition and anaemia have increased among women. Anaemia can lead to reduced productivity, greater susceptibility to infections and slow recovery from illnesses. Among women, poor nutrition can increase the risk of poor pregnancy outcomes, including obstructed labour, low birth weight babies and post-partum haemorrhage. More than one-third of married women and men are too thin, according to the body mass index (BMI), an indicator derived from height and weight measurements. Underweight is most common among the poor, the rural population, adults who have no education and scheduled castes (SCs) and scheduled tribes (STs). Overweight and obesity, the other side of malnutrition, are a growing problem in India, affecting 13 per cent of women and 9 per cent of men. Bothe these are most common in urban areas, in wealthier households and among older adults, Sikhs, and those with more education. Delhi, Kerala and Punjab have the largest percentage of overweight women and men[51].

"Pernicious anaemia", a disease caused by vitamin B_{12} defficiency, was first described by Addison in 1855 and became known as "Addison's anaemia" or "Biermer's anaemia". The symtoms of this disease include: palor, shortness of breath, jaundice, loss of body weight and muscle spasm. The cause of the disease was unknown, and it was generally fatal. In 1948 chemists Karl Folkers and Alexander Todd isolated the active component from liver extract, which they named vitamin B_{12}. The structure of vitamin B_{12} was determined by Dorothy Hodgkin in 1956, enabling new methods of production to be developed in the late 1950's. Since then, bacteria can now be used to produce vitamin B_{12} in large quantities, and pernicious anaemia has been easily diagnosed[52].

In PEM, anaemia together with multivitamin and mineral deficiency may be present. In an experiment, thew malnourished mice showed a decrease in RBCs, haemoglobin concentration and reticulocytopenia, as well as severe bone marrow and splenic atrophy. Iron (Fe) reserves in the spleen, liver and bone marrow were found to be greater in the malnourished mice. The mixed colony-forming unit assays revealed a smaller production of granulocyte-macrophage colony-forming units, erythroid burst-forming units and erythroid colony-forming units in bone marrow and spleen cells of malnourished mice[53].

Anaemia in Horse

Anaemia in horses is defined as low blood volume. There are a variety of different causes of anemia. Hence, anaemia usually occurs as secondary to some other health issues the horse may be experiencing. Anaemia can be classified as acute (rapid onset) or chronic (persistent or long-lasting). Acute anaemia is caused by rapid blood loss, such as with a traumatic injury or ruptured blood vessel. Chronic anaemia is sometimes a bit more complicated and can have many causes, some of which include cancer (especially types that affect the bone marrow), chronic kidney disease, gastric ulcers, or severe nutritional deficiencies. The symptoms of anaemia in horses are: unusually poor performance, weakness, lethargy, loss of appetite, depression, hair loses luster, heart murmur and pale mucous membranes. The diagnosis of anaemia is very straightforward, but finding the cause of the anaemia is often not as simple. A simple blood test that measures packed cell volume (PCV) will tell the veterinarian if the horse is anaemic and if so, how badly affected. The PCV does not tell the veterinarian why or how the horse became anaemic. Sometimes, further diagnostics are required to find the cause, particularly if the anemia is severe and the horse is greatly debilitated[54].

The parasites are a common concern for animals and humans who take care of them. While many are harmless, some can present serious threats to animals as well as human beings. The parasite can cause disease itself (*e.g.*, worms causing scouring and malnutrition; blood parasites causing anaemia) or parasite-borne diseases such as Lyme disease, Rocky mountain spotted fever and meningo-encephalitis can be transmitted from animals to humans (these diseases are called "zoonoses" or "zoonotic diseases"). Due to the huge amounts of potential parasites that animals might be carrying, there is a risk that people may become infected or infested from affected animals and this could be severe[55].

"Equine infectious anaemia" (EIA) is a serious disease that threatens the world's horse population. The EIA is a viral disease that affects the horse's immune system. The virus reproduces in blood cells and circulates throughout the body. The horse's immune system, via antibodies, attacks and destroys the infected RBCs. The reduced blood count causes anaemia, and associated inflammation can damage vital organs, *e.g.*, liver. Because the horse's immune system is impaired, the horse may also become susceptible to secondary infections like bronchopneumonia. The EIA-infected horses can die from the virus or from related secondary infections. Despite testing and measures to eradicate the disease, approximately 2000 new cases are identified each year in the USA. There is no cure for EIA. Although most infected horses show no symptoms, they remain infectious for life, endangering the health of other horses. EIA generally has three forms: *(1) Acute*, during which the virus is active, multiplying and harming the immune system and other organ systems; (2) *Chronic*, during which the animal may vacillate between remission and disease states, the horse may be thin or in poor body condition, and may suffer recurring bouts of clinical signs, as seen in the acute phase; *(3) Inapparent*, during which the horse carries the virus but shows no apparent symptoms, stress or disease may bring on an acute episode. This anaemia is transmitted by blood. This transmission can occur via blood-sucking insects, such as horse flies, deer flies and mosquitoes. Humans can also spread EIA in much the same way by using a single needle on multiple horses. EIA can also be passed from mare to foal in utero. The EIA is also called "swamp fever" because this disease has been associated with warm, wet regions. However, the disease is not limited by geography, and can occur anywhere there is a carrier and a vector to transmit it. EIA is difficult to diagnose because the symptoms vary from horse to horse and can mimic other diseases. In addition, some individuals may demonstrate no obvious signs. Signs may include one or more of the following: fever and/or sudden fluctuations in temperature (temperature may even exceed 105 degrees F); depression; decreased appetite; fatigue or reduced stamina; rapid breathing; sweating; rapid weight loss; bloodshot eyes with watery discharge; swelling of legs, lower chest and abdomen; weakness characterized by wobbly or rolling gait; pale or yellowish mucous membranes; irregular heart beat and/or weak pulse; colic; and abortion in mares. The only way to accurately determine whether a horse is infected with the EIA virus is by a serum test. The most commonly used method is the "Coggins test", also known as the "agar gel immunodiffusion" (AGID) test. The test was developed 35 years ago by a veterinary researcher, Dr. Leroy Coggins. The test is consistently reliable and detects the presence of EIA-specific antibodies in the blood. A positive reading indicates the horse is infected, and is a carrier of the virus. Now, a new test is coming into use, *i.e.*, the "enzyme-linked immunosorbent assay (ELISA) which offers an advantage in that the results can be assessed more quickly, especially with the "competitive ELISA" (CELISA) test. However, ELISA may not be as accurate as the Coggins test. A positive ELISA reading is verified by a standard Coggins test[56].

Anaemia in Dog

Blood is composed of a liquid portion, the plasma, and a cellular portion. That cellular portion is made up of RBCs (which carry oxygen throughout its body, platelets that allow that blood to clot when it is necessary) and WBCs (which are responsible for fighting diseases). In most anaemic pets, an underlying disease is interfering with the complex process that allows RBCs to form within the pet's bone marrow. That blood-forming process is called "erythropoiesis". There are two basic kinds of anemia. In the first kind, the pet's body loses blood faster than it can be regenerated but the animal is still able to produce new RBCs in its bone marrow. This is called *regenerative anaemia*. In the second form, the pet has lost its ability to manufacture new RBCs in its bone marrow. This type is called *non-regenerative anaemia*. A mark of regenerative anaemia is that the body is in a rush and releases some of its RBCs a bit too early. These young RBCs are called "reticulocytes". An increase in their number might alerts a veterinarian to this problem. It is sometimes called as "blood-loss anaemia[57].

Anaemia in dog can be caused by a number of different processes, including blood loss, red blood cell destruction and inadequate red blood cell production. These categories of anaemia can be caused by trauma, cancer, immune-mediated disease (a disease in which the body attacks its own cells or organs), infectious diseases, toxins, genetic defects, inflammatory diseases, iron deficiency, drug reactions, kidney failure and generalized chronic (long term) illness. Because there are so many different types and causes of anaemia, there is no gender or age predisposition for anaemia. Individual disease processes may be more common in certain age groups and breeds, so it is important to characterize the type of anaemia present. The impact of anaemia on the pet will depend on the

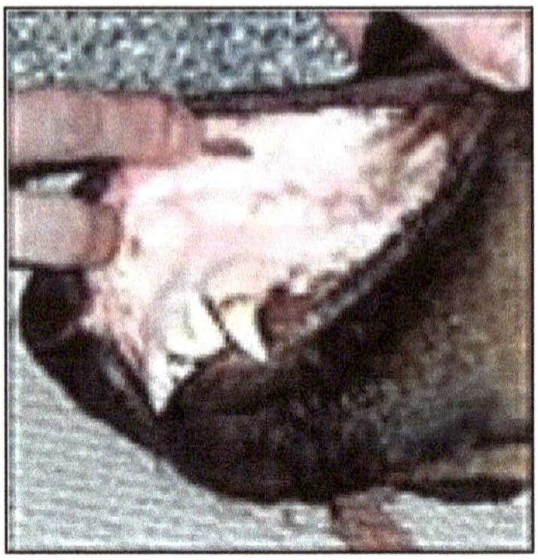

Figure 17: Severely anaemic dog showing pale gums.
[*Source*: www.petplace.com/dogs/anemia-in-dogs/page1.aspx, 2012].

cause of the anaemia as well as the severity of the anaemia. Besides this, the animals with a sudden onset of anaemia may be clinically sicker than animals with chronic anaemia. This is due to the fact that the animals can become partially adapted to the anaemia over time, and may feel relatively good in spite of the anaemia. In dogs, there may be generalized weakness, pale gums (Figure 17), increased respiratory rate, episodes of collapse, evidence of blood loss, blood in urine or faeces, external blood loss, decreased appetite, yellow discolouration of skin, abdominal distension, vomiting and weight loss. Depending on the cause of anaemia, the pet may show some or all of these signs. Some pets may not show any signs at all[58].

Anaemia in dogs occurs when they have insufficient RBC production. Since RBCs carry oxygen throughout the body, dogs with anaemia will display signs and symptoms of this decreased energy flow. Anaemia in dogs is usually a symptom of a more serious, underlying problem that can become life-threatening if left untreated. Owners should learn the signs and symptoms of anaemia in dogs to allow early detection and give their pets the best chance of surviving the condition. Two main signs of anaemia in dogs are rapid breathing and rapid heart beat without previous physical exertion. The dogs' gums are normally pink. If the gums look pale, this may be a sign of anaemia. An anaemic dog's eyes and ears may also become pale. The dogs' mood can also fade when they are not feeling well. Anaemic dogs may grow depressed and not want to participate in their usual day-to-day activities, or they may simply display overall sadness and lethargy. Flea bites are one cause of anaemia in dogs. Every time a flea bites a dog, the flea extracts blood from the dog. If multiple fleas feed on a dog, the dog can eventually lose a substantial amount of blood and become anaemic. Notable flea infestation along with other signs and symptoms should alert dog owners to possible anaemia. Another sign of anaemia in dogs is generalized weakness. An anaemic dog may not have the strength to run as fast or as long as usual, play for the normal length of time it usually plays, or pull and push toys as robustly as it usually does. Also, a dog's appetite is usually one of its most consistent characteristics. If the appetite shows a marked decrease, the dog may be anaemic. Dogs with anaemia may also groom less, or stop grooming themselves altogether. The anaemic dogs may sleep for far longer periods of time than they normally do. A dog that has experienced blood loss may become anaemic, so a dog that has been in an accident and has sustained blood loss should be examined by a veterinarian. An anaemic dog may also collapse because of the oxygen deprivation the dog's body is experiencing. A dog's abdominal area should be proportionate to the rest of its body. If a dogss abdominal area becomes abnormally larger than its normal size, the dog may be anaemic. The owners should examine their dogs' urine and faeces regularly to spot signs of trouble. An anaemic dog may pass blood in its urine and/or faeces[59].

It has been further stated that the adult dogs become anaemic when the concentration of RBCs in whole blood is less than 37 per cent by volume. The normal range is 39 per cent to 60 per cent. The RBCs are produced by the bone marrow and have an average life span of 110 to 120 days. Old RBCs are trapped by the spleen and removed from the circulation. The iron they contain is recycled to make new erythrocytes (RBCs). The purpose of RBCs is to carry oxygen. Thus, the symptoms of anaemia are caused by insufficient oxygen in the organs and muscles. Signs include lack of appetite, lethargy and weakness. The mucous membranes of the gums and tongue become pale pink to white. In dogs with severe anaemia, the pulse and respiratory rate are rapid, and the dog may collapse with exertion. A heart murmur may be heard. There are many conditions in which a watery or mucus discharge overflows the eyes and runs down the face. With a severe watery eye, there is constant wetness and the skin may become inflamed and infected, adding to the dog's unsightly appearance and physical discomfort. Epiphora is primarily a cosmetic problem unless it causes inflammation, or is a symptom of a painful eye. For example, entropion, conjunctivitis, foreign bodies, corneal ulcers, anterior uveitis and acute glaucoma are all accompanied. Thus, the anaemia in dogs can be caused by blood loss, haemolysis, or inadequate RBC production[60].

"Blood-loss anaemia" in adult dogs is most commonly caused by trauma, slow gastrointestinal bleeding associated with stomach and duodenal ulcers, parasites, and tumours in the gastrointestinal tract. Chronic blood loss also occurs through the urinary system. Hookworms and fleas are common causes of chronic blood loss in puppies. In "haemolytic anaemia", haemolysis occurs due to acceleration in the normal process of RBC breakdown. The RBCs break down to form bile and haemoglobin. With severe haemolysis, these breakdown products accumulate in the body. Accordingly, in a dog experiencing an acute haemolytic crisis, jaundice and haemoglobinuria (passing dark-brown urine that contains haemoglobin) are seen. Additionally, the dog appears weak and pale, and has a rapid pulse. The spleen, liver and lymph nodes may be enlarged. Causes of haemolysis include immune-mediated haemolytic anaemia, congenital haemolytic anaemia, infectious diseases (*e.g.*, canine babesiosis and leptospirosis), drug reactions to medications such as acetaminophen and poisonous snake bites. A number of bacteria produce toxins that destroy RBCs, so haemolysis can also occur with severe infections. "Immune-mediated haemolytic anaemia" is the most common cause of haemolysis in adult dogs. The RBC destruction is caused by auto-antibodies that attack antigens present on the surface of the cells, or by antigens from medications or organisms attached to the RBC walls. The weakened cells are trapped in the spleen and destroyed. Poodles, Old English Sheepdogs, Irish Setters and Cocker Spaniels are predisposed to immune-mediated haemolytic anaemia, but all breeds are susceptible. The affected dogs are usually between 2 and 8 years of age. Most cases of immune-mediated haemolytic anemia are idiopathic. That is, the reason why the auto-antibodies developed in that particular dog is unknown. In some cases, there is a history of recent drug therapy. An immune-mediated haemolytic anaemia also occurs with systemic lupus erythematosus. The diagnosis is made by microscopic examination of blood smears, looking for specific changes in the appearance of the erythrocytes and other blood elements; and by serologic blood tests. In "congenital haemolytic anaemia", many inherited abnormalities in the structure of RBCs can result in their premature destruction. "Phosphofructokinase deficiency" is an autosomal recessive trait that occurs in English Springer Spaniels and Cocker Spaniels. A deficiency of this enzyme results in changes in the pH of RBCs, causing the cells to periodically fragment and produce bouts of haemoglobinuria. There is no effective treatment of this condition. "Pyruvate kinase deficiency" is another RBC enzyme deficiency caused by an autosomal recessive gene. This disease is recognized in many breeds, including Basenjis, Beagles and West Highland White Terriers. The puppies usually develop the haemolytic anaemia at 2 to 12 months of age. Death by 3 years of age is the usual outcome[60].

Anaemia in Cat

Cats are particularly prone to the development of anaemia and hence we often see anaemic feline patients. Feline RBCs have a shorter lifespan (70 days) than many other mammalian species (*e.g.*, 110-120 days in dogs and humans), meaning that anaemia can develop quite rapidly in cats. In regenerative anaemias, haemorrhage can arise due to injury in road traffic accidents, internal bleeding due to tumours, severe parasitism with fleas and lice (especially in kittens) and loss of blood from the gastrointestinal tract or urinary tract because of disease of these systems. Haemolysis may be due to: (a) infections, *e.g.*, feline leukaemia virus (FeLV), feline *Haemoplasma* infection; (b) secondary to drugs, tumours, very low blood phosphate concentrations; (c) Heinz body haemolytic anaemia due to oxidative injury to feline RBCs (can be seen with onion poisoning from feeding baby foods with high onion levels, paracetamol toxicity and disease states such as diabetic ketoacidosis, hyperthyroidism and lymphoma); (d) incompatible blood transfusions; (e) neonatal isoerythrolysis can occur when type A kittens are born to type B queens, and they ingest antibodies in the colostrum that destroy their RBCs resulting in anaemia- this is one cause of fading kittens; (f) inherited defects (rare) such as osmotic fragility (reported in Abyssinians and Somalis), porphyria (in Siamese, also causes discolouration of teeth), pyruvate kinase deficiency (in Abyssinians and Somalis). In cats, most anaemias are non-regenerative. In this anaemia- (a) primary bone marrow disorders include diseases known as pure red cell aplasia, pancytopenia (which can arise due to FeLV, FIV or parvovirus infection), abnormal maturation of bone marrow cell lines and leukaemia; (b) systemic causes of bone marrow suppression such as chronic renal failure (where the kidneys fail to produce adequate amounts of erythropoietin which stimulates RBC production in the bone marrow), anaemia of chronic disease (many diseases cause secondary reduction in RBC production leading to mild anaemia) and FeLV or feline immunodeficiency virus (FIV)- associated non-regenerative anaemia. In addition, causes of anaemia can often be present at the same time in one cat, posing a diagnostic challenge in the investigation of anaemia, *e.g.*, haemolysis caused by *Haemoplasma* species usually results in a regenerative anaemia, but concurrent FIV infection can inhibit the RBC regenerative response resulting in a non-regenerative anaemia. A common sign of anaemia in cats is paleness of the mucous membranes, such as the gums (Figure 18). Cats with severe anaemia can show weakness, an increased heart rate and increased respiratory rate. Anaemic cats may develop pica (a craving for unusual foods) which is typically manifested as licking concrete, eating cat litter or soil. Occasionally cats become jaundiced (yellow discolouration of mucous membranes) with severe acute RBC breakdown (haemolysis). Cats may also show signs of underlying diseases causing anaemia such as chronic renal failure or feline infectious peritonitis (FIP). Cats which having a rapid fall in RBCs are more likely to show severe clinical signs of anaemia than those cats with a slowly developing chronic anaemia[49].

Anaemia is detected by finding reduced RBC numbers in a blood sample collected from the patient. In feline anaemia, the cats have a

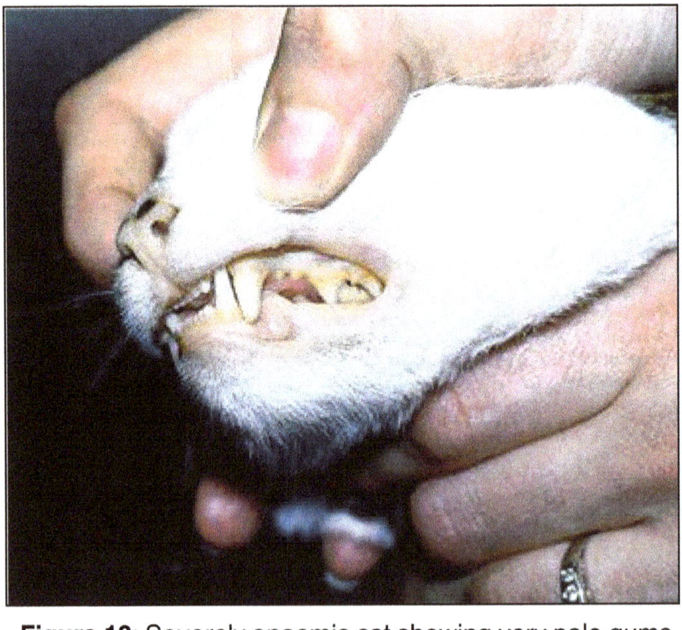

Figure 18: Severely anaemic cat showing very pale gums.
[*Source*: http://fabcats.org/owners/anaemia/info1.html, 2012].

reduction in RBC count, haemoglobin concentration and PCV (obtained by spinning a small blood sample in a glass tube and observing the percentage volume taken up by the centrifuged RBCs). Next, determine whether the anaemia is regenerative or not. This is done by evaluating the cat's haematological parameters as well as observation of a stained blood smear. Haematological changes reflecting regeneration include variation in RBC size and the presence of large immature RBCs. The number of reticulocytes (immature RBCs) present in the blood indicates whether active RBCs production is occurring in the bone marrow, and is used to definitively determine whether an anaemia is regenerative (with lots of reticulocytes) or not (few or no reticulocytes). As both feline leukaemia virus (FeLV) and FIV can cause anaemia, all anaemic cats should be screened for these viruses. Other tests for infectious diseases may be required, depending on the patient's clinical signs, such as for feline parvovirus or feline *Haemoplasma* infection (feline infectious anaemia, previously known as *Haemobartonella felis*). When systemic diseases, such as chronic renal failure, are believed to be the cause of anaemia a thorough investigation of different organ systems is required. This could include biochemical testing on blood and urine samples, radiography, ultrasonography etc. When a bone marrow problem is suspected, collection of bone marrow samples may be required to establish a diagnosis. These samples can be collected from either the forelimb (often the humerus) or hind limb (from the shaft of the femur). A general anaesthetic is required for this procedure in cats[49].

Furthermore, the anaemia is common in critically ill cats. The origin may be either acute or chronic. The diagnostic workup for cats with acute anaemia should include: evaluation of blood smear, slide agglutination test, complete blood count, new methylene blue stain for Heinz bodies, Coomb's test, blood typing, FeLV/FIV testing, chemistry panel, coagulation testing (if evidence of bleeding) and PCR for haemotropic mycoplasmas. The causes of "acute anaemia" include haemorrhage or haemolysis. Cats are more likely to develop anaemia following blood loss than dogs because they have a reduced blood volume (60 ml/kg versus 90 ml/kg). In addition, the cats are more at risk for toxic anaemias because they are predisposed to Heinz body anaemia because of extra sulphydril groups on feline haemoglobin that are prone to oxidative injury. In "parasitic anaemia", the blood smears from anaemic cats should be evaluated for RBC parasites. *Mycoplasma hemofelis* (formerly *Hemobartonella felis*) causes a regenerative haemolytic anaemia that is usually positive on the direct antiglobulin test. The organisms appear as chains of small cocci on the RBC surface. "Toxic anaemia" is caused by oxidants such as zinc, acetaminophen (paracetamol), local anesthetics, new methylene blue, propofol, propylene glycol and onion powder in baby food. Acetaminophen causes methaemoglobinaemia as well as Heinz body anaemia. The classic signs of these anaemias include brown coloured blood, 'muddy' mucous membranes, facial oedema, tachypnoea and weakness. Primary "immune mediated haemolytic anaemia" is rare in cats, so secondary causes such as FeLV, FIV, blood parasites and drugs should be ruled out. Cats with immune mediated haemolytic anaemia rarely develop pulmonary thromboembolism, and may have a better prognosis than dogs if they are negative for FeLV and FIV. "Hypophosphataemia" may cause haemolytic anaemia in cats with diabetes, hepatic lipidosis, or malnutrition when nutrition is reintroduced. After starvation, when carbohydrates are given, phosphorus is consumed in the production of ATP and enzymes in the glycolysis pathway, resulting in hypophosphataemia. Reduced phosphorus can lead to haemolytic anaemia, muscle weakness. Causes of "chronic non-regenerative anaemia" include feline leukaemia, feline immunodeficiency virus, myelodysplasia, infiltrative neoplasia, chronic blood loss, chronic renal failure or anaemia of chronic disease. Diagnostic testing should be performed to rule out FeLV and FIV in all anaemic cats. Additional work-up should include complete blood count, serum chemistries and urinalysis to rule out extra-marrow disease such as chronic infection, liver disease or renal disease. Bone marrow evaluation is generally indicated in cases of pancytopenia or non-regenerative anaemia with no evidence of extra-marrow disease[61].

Once again, it is important to elucidate the main causes of anaemia which include: regenerative cause; haemorrhage which may be obvious such as a laceration, or less obvious such as internal bleeding; parasites such as worms, fleas or blood parasites (*Haemobartonella* or *Cytauxzoon*); immune-mediated haemolytic anemia (IMHA), a disease in which the cat's own immune system can become directed against its own RBCs; Heinz-body haemolytic anaemia; ingestion of certain medications, foods or toxins *e.g.*, acetaminophen, naphthalene, mothballs, propylene glycol, copper, zinc, onion, etc.; neonatal isoerythrolysis; non-regenerative; FeLV; FIV; chronic renal failure; some cancers; poor nutrition (starvation); and chronic inflammatory diseases. The main signs of anaemia in cats are: pale mucous membranes; Lethargy; loss of appetite; weight loss; generalized weakness; rapid breathing and pulse (tachypnoea); heart murmur; sleeping more than usual; and black, tarry stools. The veterinarian will perform a complete physical examination of the affected cat and obtain a medical history from the owner. For diagnosis, the veterinarian will have to perform: (a) Complete blood count a series of tests which evaluates the cellular components of blood (RBCs, WBCs and platelets) and the elevated numbers of reticulocytes (immature RBCs) would point to regenerative anaemia; (b) Blood smear to check for the presence of blood parasites and abnormalities of RBCs; (c) Biochemical profile to evaluate the overall health of the cat and determine if there is any organ involvement; (d) Faecal examination if gastrointestinal blood loss is suspected; (e) X ray to check for foreign objects, organ size, tumours, etc.; and (f) Bone marrow biopsy for cats with non-regenerative anaemia to check for Coombs test (also known as antiglobulin test or direct antibody test), this test is to detect the presence of antibodies which can bind to the surface of RBCs[62].

Some Animals Cases of Malnutritional Anaemia

Generally, a backyard butcher does not feed his animals, does not provide clean water or adequate shelter and does not give medical attention when they are sick or injured. In the picture given below (Figures 19-21[63]), the animals (horse and cattle) have become emaciated, malnourished, scared to death, they have infections, open sores, are infested with parasites, their eyes are infected causing blindness, their noses are runny and they have hacking coughs. They have high fevers and the horse has a huge sore on her leg that is infected down to the bone. These animals are very sick.

In the Figures 22–23, the goats are seen from severely abused and neglected. They are all riddled with lice, parasites and ringworm; have eye infections; are malnourished, emaciated, sick, weak and terrified of humans.

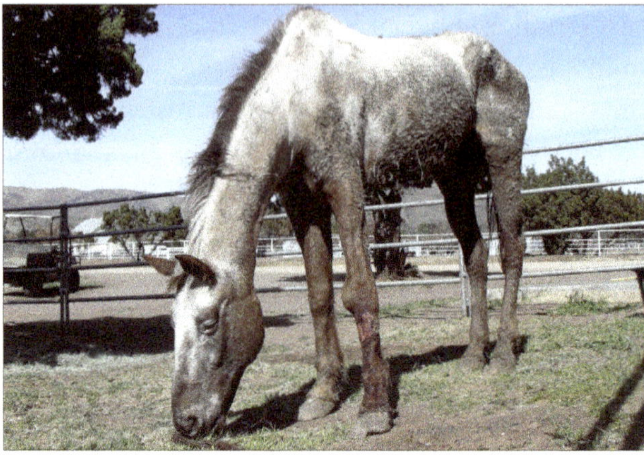

Figure 19: Horse at top; **Figure 20**: Cow at middle and **Figure 21**: Calf at bottom: All malnourished and anaemic animals.

[*Source*: The Gentle Barn, Santa Clarita, CA; Website, 2012].

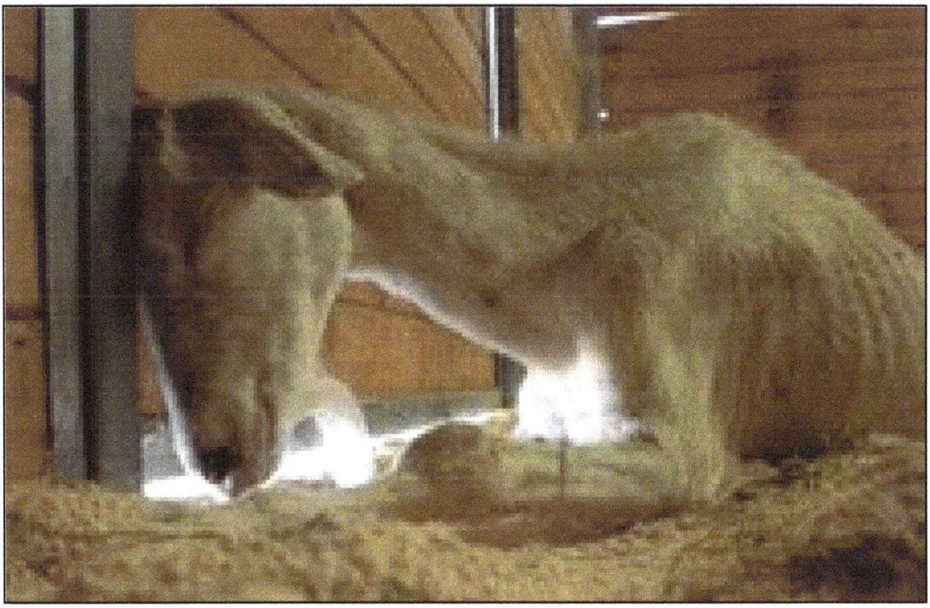

Figure 22: At top and **Figure 23**: At bottom:
All emaciated, infected, malnourished and anaemic goats.
[*Source*: The Gentle Barn, Santa Clarita, CA; Website, 2012].

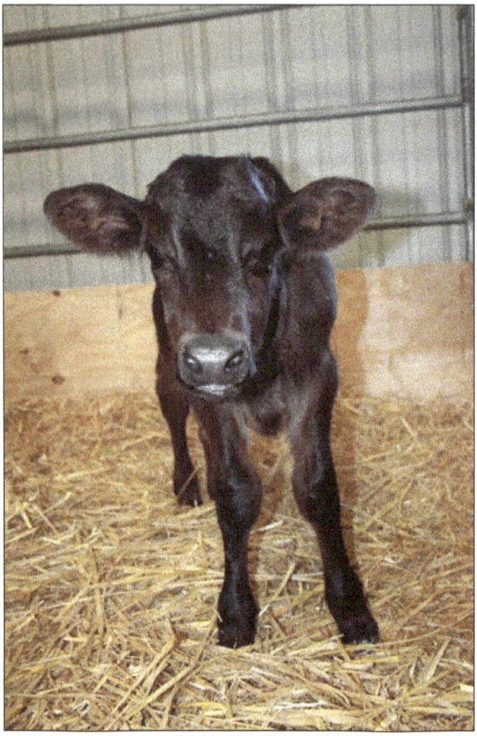

Figure 24: At top and **Figure 25**: At bottom:
All sick, emaciated, infected, malnourished calves.
[*Source*: The Gentle Barn, Santa Clarita, CA; Website, 2012].

The Figures 24–25 of calves showing that the the calves are filthy, emaciated, sick, weak and afraid. They also have infections, wracking cough and high fever for a long time.

4
Special Bulletin for Malnutritional Cancer in Animals

Dog and Cat

"Malnutrition" is very common in sick animals and is thought to affect up to 50 per cent of hospitalized dogs and cats. Within days malnutrition can begin to have deleterious effects on immune function, healing and, most importantly, quality of life. While human patients will follow their doctor's instruction to eat even though they don't feel well or have a good appetite, it is impossible to convince a sick pet to eat if they don't feel like it[20,64] (Figures 26–27[64]).

Due to malnutrition, different types of cancer may develop in dogs (Figures 28-29[65]). Some symptoms of dog cancer are rather indistinct and, especially at the onset of the disease, it is difficult to tell whether a dog showing some cancer-like symptoms is in fact suffering from canine cancer or some other illnesses. Some cancers (*e.g.*, lymphomas, mast cell tumors, fibrosarcomas) can arise as lumps and bumps on the dog's skin. Fortunately not all bumps and lumps are cancerous. Very often, benign growths such as warts, fatty tumours, cysts, etc. can arise in a dog's skin as well. Usually, any lumps and bumps that appear, then decrease in size or even disappear, and finally reappear in a larger size are more likely to be cancer than benign lumps. However, the only way to tell whether a lump is cancerous

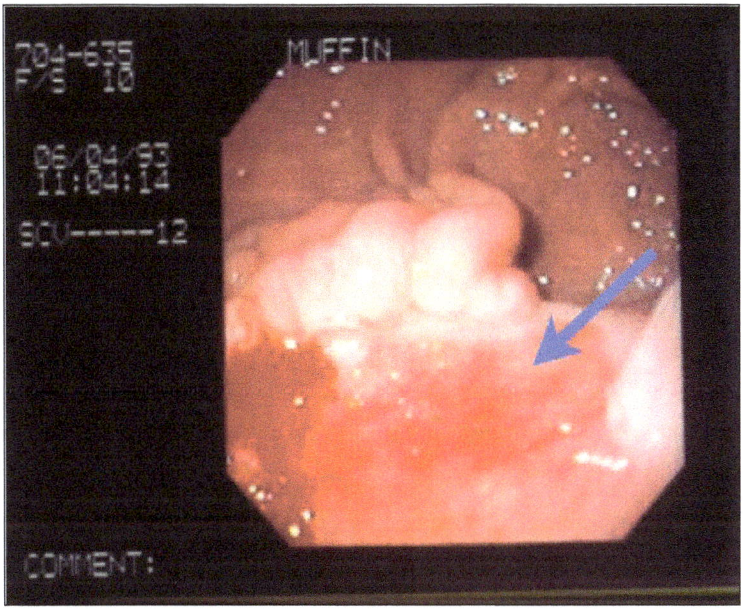

Figure 26: Rectal tumour in dog seen on colonoscopy (arrow).

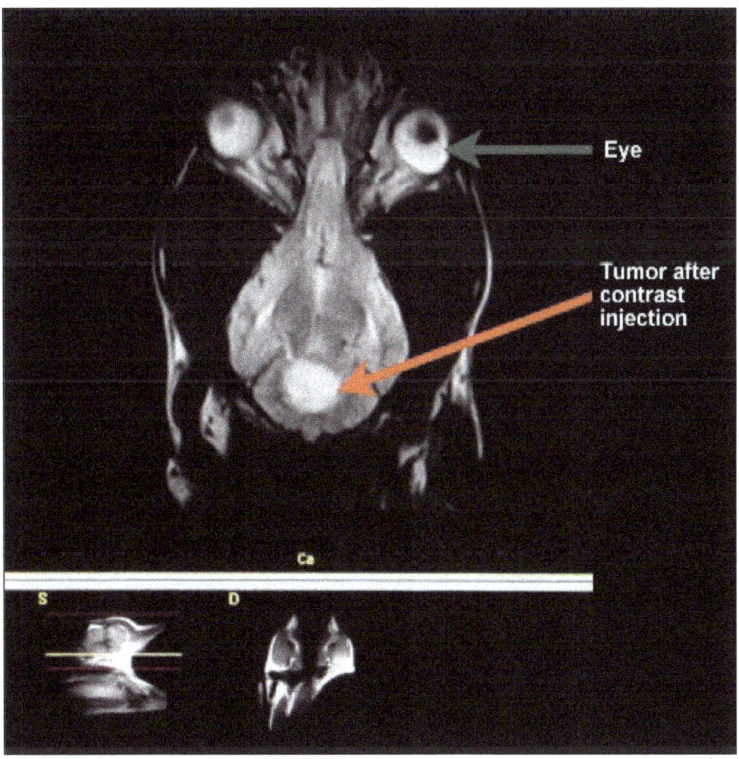

Figure 27: Brain (cerebellum) tumour in dog seen on magnetic resonance imaging (MRI).

[*Source*: Animal Specialty Hospital, Rockledge Florida.htm., 2011].

Figures 28–29: Symptoms of cancer in dogs.
[*Source*: Natural-Dog-Health-Remedies.com, 2011].

or not is by a quick and painless process called fine needle aspiration. While coughing and sneezing are common symptoms of dog cancer, there are also quite a few other dog illnesses which have the same symptoms. Acute coughing is usually caused by some kind of allergies or infection (*e.g.*, kennel cough), while chronic coughing may be caused by heartworm infection, fungal infection of the lung, or heart disease. Sneezing can also be the result of allergies. However, if the sneezing is accompanied by a bloody nasal discharge from only one side of the nasal cavity, and it has been ruled out that there are no foreign bodies (*e.g.*, foxtail) trapped up the dog's nasal cavity, then cancer should be suspected. Chronic disorders of the gastrointestinal tract such as vomiting and diarrhea can be symptoms of dog cancer of the gastrointestinal tract, or cancer anywhere else. There are, however, several dog illnesses that can cause the same symptoms. The most common one is inflammatory bowel disease, and another possibility is Addison's disease. It is important to have an early diagnosis of whatever disease the dog is suffering, since chronic inflammation of the gastrointestinal tract may eventually progress to cancer. Appetite loss is also one of those symptoms of dog cancer that is indistinct, as many other diseases (*e.g.*, dental or gum diseases, inflammatory bowel disease, heart disease, kidney disease, etc.) can cause appetite loss in dogs as well. Also, older dogs tend to eat less. Chronic skin diseases such as skin sores, itch and lesions, and wounds that do not heal, can be indicative of canine skin cancer. However, bacterial or fungal infections, or immune disorders can also cause these symptoms. Blood in urine or urinary incontinence can be caused by some types of dog cancer (*e.g.*, bladder cancer), but they can also be caused by urinary tract infections, bladder stones, kidney disease, diabetes, Cushing's disease, etc. While cancer can cause a lot of pain to a dog, so can many other diseases, such as arthritis, ear infections, dental or gum diseases, etc. Bone cancer in dogs can cause lameness. However, lameness can also be caused by arthritis, hip dysplasia, or some form of injuries. Perhaps lethargy and weakness are one of the most vague symptoms of all diseases. A dog can be lethargic due to a various reasons, *e.g.*, pain from arthritis, anemia, fever caused by infections, etc. Oral cancer in dogs can cause bad breath, but perhaps more frequently, bad breath is caused by indigestion, dental/gum diseases, liver disease and kidney disease. Brain tumours in dogs can cause seizure, but it can also be seen in thyroid problems, kidney disease, liver disease, hypoglycaemia and poisoning. Many of the above symptoms of dog cancer are not unique to cancer. That is why very often cancer is not diagnosed until it is at a later stage. Finally, whenever a dog has a disorder that does not respond well to treatment in one to two months, and is showing some symptoms of dog cancer, then the possibility of cancer should be considered and a more thorough examination should be done[20,65].

Horse and Donkey

In the so called 'Third World', where poverty is common, millions of horses and donkeys work day in day out as cart-horses, pack-mules or as horse for a carriage. The animals don't get good food and don't get enough to drink. They have to work in the burning sun, on busy and dusty roads without one moment of rest. The owners of the animals are often too poor to go to a 'Vet' with their sick or wounded animals. Next to that some owners don't even know that an animal feels pain. A lot of animals have to work till they drop dead, and that often happens in the streets (Figure 30[66]). The owners of the horses and donkeys are often very poor. The owner really needs the money, because he has to take care of his family. When the animal becomes sick, the owner is not able to make money because he doesn't have the horse or donkey to pull. In malnutrition, the horses grow out of shape. Because of all the heavy work, the horses can not develop physically; often legs and ankles become crooked. They have often a bad back also, since they have to carry a too heavy load and because

Figure 30: Hard-working horses.

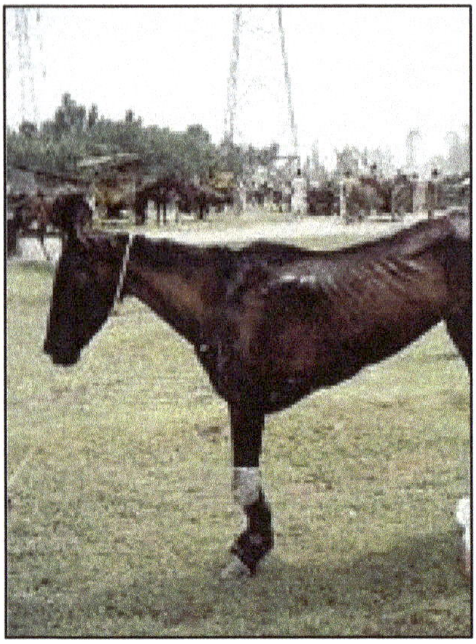

Figure 31: Horse showing crooked leg, bad back, over tireness and emaciation due to heavy work.

Figure 32: Donkeys carrying heavy materials.

[*Source*: Animal World-Brook Hospital for Animals.htm, 2011].

it is loaded in a wrong manner. The animals have to work in the burning sun with terribly high temperature. The owners do not give them enough to drink so they dehydrate. Due to less consumption of water and food the animals are exhausted and overly tired (Figure 31[66]). Through malnutrition, starvation, dehydration and exhaustion occur. In the Netherlands, the horses are generally used for riding. In the beginning of the 1900's there were no cars in the streets, but carriages, horse and cart. This is what in Egypt, Jordan, Pakistan and India too. In these countries, the cars are expensive so horses pull the carriages that transport people;

and horses and donkeys have to pull carts with heavy loads like building materials (Figure 32[66]). These animals are seen as a kind of a 'motor' to do the heavy works. In these countries too, it is usually very hot and we cannot cope with that, but the horses and donkeys have a problem with it also. In World War I (1914-18), the British army was present in Egypt. In the beginning of this war, about 20,000 horses and donkeys from the UK were brought ashore. When the war ended in 1918, the horses and donkeys were sold to Egyptian people so the poor animals were left there[20,66].

5
Malnutritional Cancer Remedies by Edible Plants

General Consideration

Though many cancers can be controlled, the chronicity of the disease and expenses on supportive treatment are the valid reasons of not getting the cancer cured. Recent researches revolve round the urgency to evolve suitable chemotherapy consistent with new discoveries in cell biology for the treatment of cancer with no toxic effect on normal cells. Toxicological aspects of drug warrant serious consideration without which the therapeutic concept of clinical efficacy will remain misconceived. Chemotherapy being a major treatment modality used for the control of advanced stages of malignancies and as a prophylactic against metastasis, exhibits severe toxicity on normal tissues[2–3,6,8–9,20,25,31].

If malnutrition results in cancer, it is very difficult to treat, but some improvements can be brought about by taking proper foods or nutrients and drugs[3,8]. Treatment for malnutrition depends on the cause. The treatment may include increasing the calories and nutrients in the diet by doing one or more of the followings:

1. Eating large meals and snacks between meals;
2. Eating small meals often throughout the day if the patient has trouble in eating large meals;

3. Eating or drinking a nutrition supplement if the patient has trouble in eating the right kinds and amounts of food.

Vitamin or mineral supplements are needed if the body is low in vitamins and minerals. Medicines may also be given if health problem appears due to malnutrition. If buying or preparing the right kind of foods is not possible, then informations about community programmes can help. A diet for a patient suffering from specific disease is called a "therapeutic diet". Thus, the therapeutic diets are modified versions of the normal or regular diets. Diet therapy helps patients to lead a full life. Without it the disease can become uncontrollable irrespective of whether the patient is also prescribed drugs or not. Various types of dietary modifications that become necessary include: change in consistency (*e.g.*, fluid and soft diets, increase or decrease in energy content or contribution (*e.g.*, low and high caloric diets), inclusion of more or less amounts of one or more nutrients or other substances, increase or decrease in fibre content, elimination of spices and condiments (*e.g.*, bland diets), inclusion or exclusion of specific foods and change in intervals of feeding, *i.e.*, meal frequency. Mild form of a nutrient deficiency would be treated and controlled by eating foods rich in that particular nutrient. In addition, tablets, capsules or syrups which contain the nutrients in concentrated form may be required especially in cases where the condition is moderate to severe. These facts suggest that nutrients must be supplied to the body in the right amounts and proportions for a person to remain healthy. If the diet lacks or is deficient in a particular nutrient, the body will also become deficient in that nutrient. When this deficiency is prolonged or sufficiently severe, the person starts showing signs of nutritional deficiency disorders[3,19].

To prevent energy deficiency, carbohydrate-rich foods (*e.g.*, sugars, cereals, roots and tubers, fruits like banana, sapota and mango) and fat-rich foods (*e.g.*, nuts, oilseeds, fish, meat, vegetable oil, ghee, vanaspati and butter) should be consumed. If protein is lacking in the diet, pulses, milk and milk products, eggs, meat, fish, nuts and oilseeds should be taken. If vitamins and minerals are deficient in the diet, green leafy vegetables, nuts, pulses, eggs, meat, butter, milk, liver, fish, etc. are to be eaten. In addition, fibre also has a valuable role to play in the prevention of diseases such as constipation and cancer of the colon. Some complex carbohydrates (*e.g.*, cellulose) and non-carbohydrate forms of fibre are available, which may be helpful in treating the colon cancer. In PEM, high energy foods, oral rehydration solution (ORS) in the case of mild to moderate diarrhoea, a high energy-high protein diet when tolerance improves for solid foods, and mineral and vitamin supplements, etc. are given. In chronic diseases like cancer in which tissue wasting takes place, a high caloric-high protein diet is best. For this purpose, a suitable dietary guideline follows as: cereals–400 g, pulses–50 g, roots and tubers–100 g, green leafy vegetable–200 g, other vegetables–200 g, eggs/paneer–2/60 g, milk and milk products–1 litre, fats and oils–25 g, sugar–50 g and tea/coffee–7/15 g[3,8,19].

A nutrition programme is a programme planned to combat malnutrition in the country[3,34]. Hence, various nutrition programmes have been launched by the Government of India in order to meet the needs of the community. These programmes can be mainly classified into two categories:

(A) Food Supplementation Programmes/Nutrient Deficiency Control Programmes

(B) Nutrition Education Programmes

The nutrition programmes are:

1. Applied Nutrition Programme (ANP)
2. Special Nutrition Programme (SNP)
3. Composite Nutrition Programme (CNP)
4. Mid-Day Meal Programme (MDMP)
5. Integrated Child Development Services (ICDS)
6. Oxfam Project-Nutrition Rehabilitation (OPNR)
7. Nutrition Health Education and Environmental Sanitation (NHEES)

It is now generally recognized that the most rewarding strategy for combating malnutrition will be an integrated programme which will necessarily include nutrition education. All these nutrition programmes may be helpful in ameliorating various cancers caused by malnutrition.

Many factors contribute to the development of eating disorders, and because each individual's situation is different, the "best treatment" must be tailored for that individual. The process begins with an evaluation by a physician or psychiatrist. Many patients receiving long-term dialysis may have received suboptimal dialysis, which may result in uremic symptoms such as lethargy, drowsiness, nausea, vomiting and anorexia. Frequent monitoring of the delivered dialysis, urea removal and plasma urea levels are needed to minimize the development of uremia and risk of malnutrition[3].

Malnutrition affects a high number of hospitalized patients. Nutritional support is then essential and in most of the cases dietary supplementation or enteral feeding is given. Parenteral nutrition may be required in some cases of severe malnutrition. Parenteral nutrition provides an adequate nutritional support when oral or enteral nutrition is impossible or insufficient. The energy supply by a mixture of nutrients in multi-compartment bags is now routinely accepted and all-in-one mixtures facilitate home parenteral nutrition. The benefits of this therapy are obtained with a trained team but the most undernourished patients should not be deprived of its advantages[3,67].

Hospitalization may be needed for those individuals who exhibit: (a) a significant weight loss (more than 30 per cent less than normal weight), particularly if the weight loss has been recent and rapid; (b) medical complications such as seizures, heart failure, or irregular heart rhythms; (c) chemical imbalances or dehydration; (d) an overriding psychiatric problem such as depression or thoughts of suicide; and (e) no response to outpatient treatment after 3-4 months. The goals of outpatient treatment of eating disorders (malnutritional cancer) are: nutritional rehabilitation (balanced diet), weight restoration, stopping weight decreasing behaviours, improvement in eating behaviours, and improvement of psychological and emotional states. The goals for hospitalized individuals are the same as outpatient management but with increased intensity. If a patient is admitted to the hospital for treatment, resolving medical complications and stabilizing nutritional status are the first and most important goals. If severe weight loss and malnutrition are apparent, intravenous feeding will be needed[3].

Consideration of many factors may be related to decreased appetite and/or involuntary weight loss. All health professionals who deal with elderly residents of long-term care facilities should be made aware of the dangers of involuntary weight loss and the fact that early interventions greatly improve the chances for a successful outcome. Family therapy is important for patients who live at home because family dynamics play an important role in eating disorders. Parents and siblings can be deeply affected by the presence of an eating disorder within the family and need an outlet to understand the disease and recovery process. Family therapy provides a meeting place to communicate concerns and needs between the family and the patient. Group therapy, when the individual is ready, can be an important source for peer support. Goals of group therapy typically include: (a) exploring underlying emotional conflicts that may be expressed by eating behaviours; (b) sharing problem solving and effective coping strategies; (c) developing realistic weight goals and a healthy relationship with food; and (d) improving interpersonal communications. In nutritional therapy, a professional nutritionist or dietician can help patients learn how to manage their weight effectively. Individualized guidance and a meal plan that provides a framework for meals and food choices (but not a rigid diet) are helpful for most individuals. Nutritionists can also help to have better how their eating disorders can create serious medical problems. Drug therapy should be used in combination with psychotherapy and nutritional therapy. The success of therapy often depends on the individual's drive to seek help. Seeking treatment from physicians and psychiatrists offers the greatest success in the recovery[3].

"Our mothers were right when you're sick you have to eat more nutritious diet to get well". Even more importantly, we now recognize that proper nutrition is critically important for the prevention and management of early disease. The same is true in our pets/animlas. Therefore, for the treatment of malnutritional cancer and for other diseases too, not only the drug therapy is required but also dietary (nutritional) treatment is essential. The best nutritions for the animal patients as well should be instituted early. Malnutrition is very common in sick animals and is thought to affect up to 50 per cent of hospitalized dogs and cats. Within days malnutrition can begin to have deleterious effects on immune function, healing and, most importantly, quality of life. While human patients will follow their doctor's instruction to eat even though they don't feel well or have a good appetite, it is impossible to convince a sick pet to eat if they don't feel like it. Options for nutritional support include the use of appetite stimulants to get the patients eating again, short-term feeding tubes when appetite stimulants are not effective, or long-term feeding tubes for those patients that will require prolonged nutritional support to recover from serious illness. The use of intravenous (i.v.) nutrition (not just put glucose in i.v. fluids) in animals is also recommended to provide

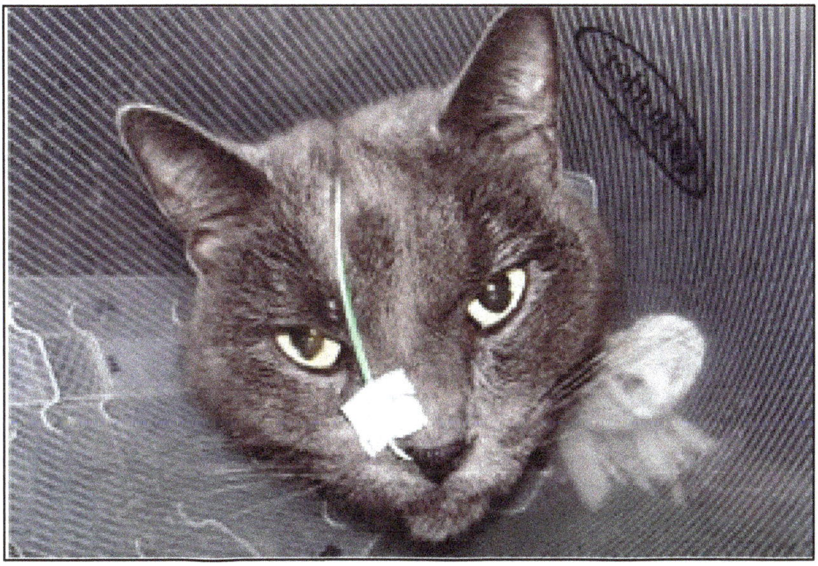

Figure 33: Cat with a nasoesophageal tube in place.

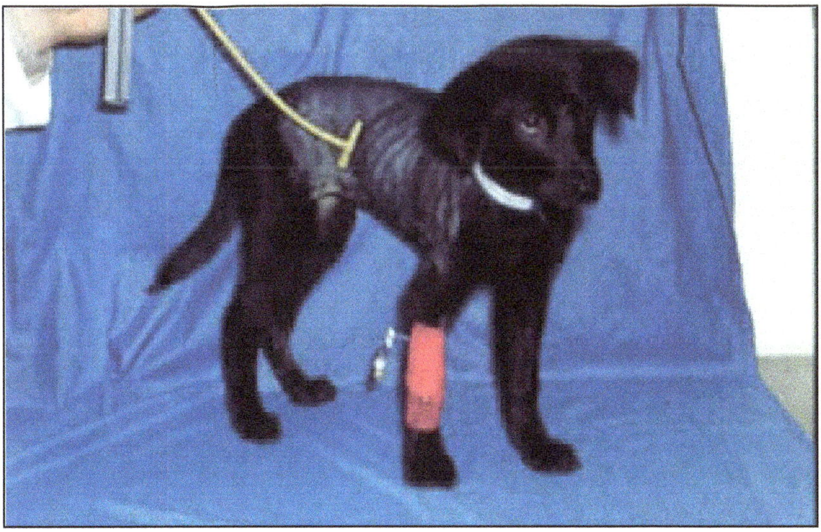

Figure 34: Gastrostomy tube in place in a puppy.
[*Source*: Animal Specialty Hospital, Rockledge Florida.htm., 2011].

complete nutritional support for recovery of our patients who do not have a functional digestive tract[20,64] (Figures 33–35[8]).

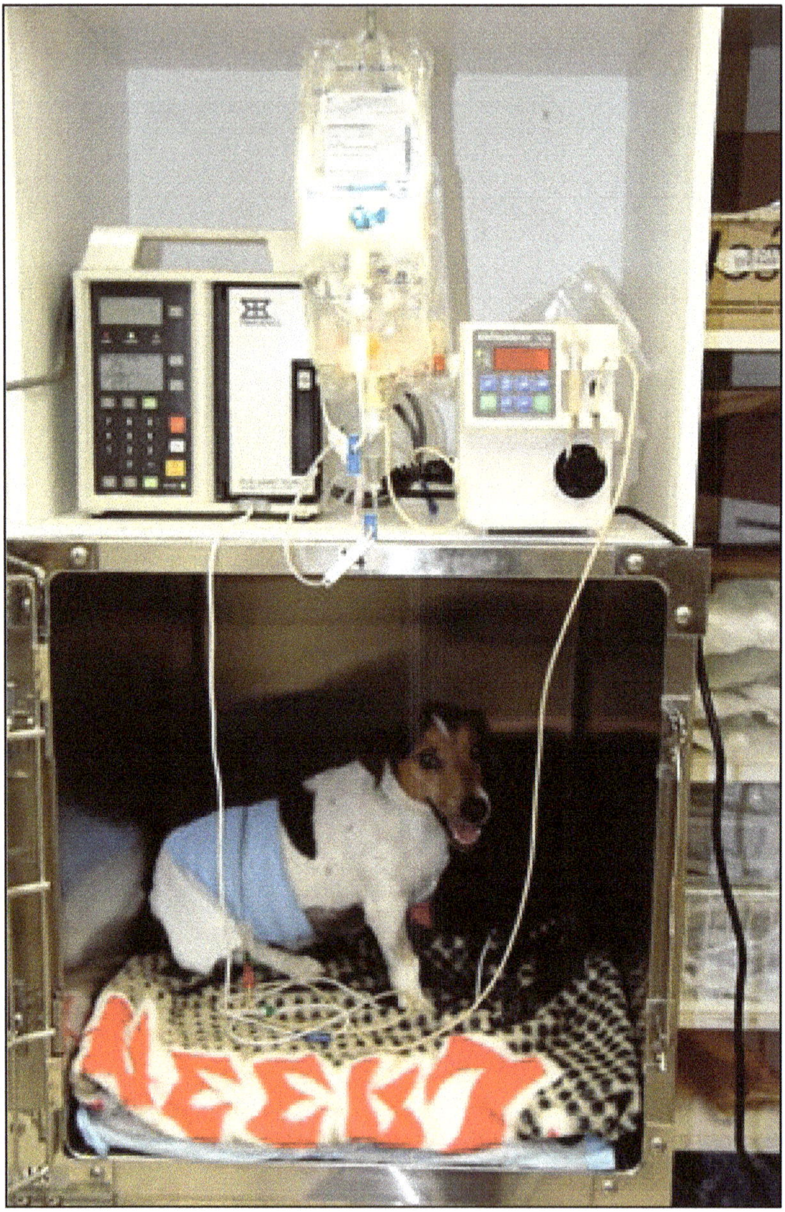

Figure 35: Patient (Dog) receiving i.v. partial parenteral nutrition. [*Source*: Animal Specialty Hospital, Rockledge Florida.htm., 2011].

Actions of Edible Plants (Vegetables and Fruits) against Malnutritional Cancer

Many edible plants and their products, particularly vegetables and fruits (Figures 36-44) exhibit potent anticancer activity against several cancers. Consumption of large amount of fruits and vegetables can prevent

Figure 36

Figure 37

Figure 38

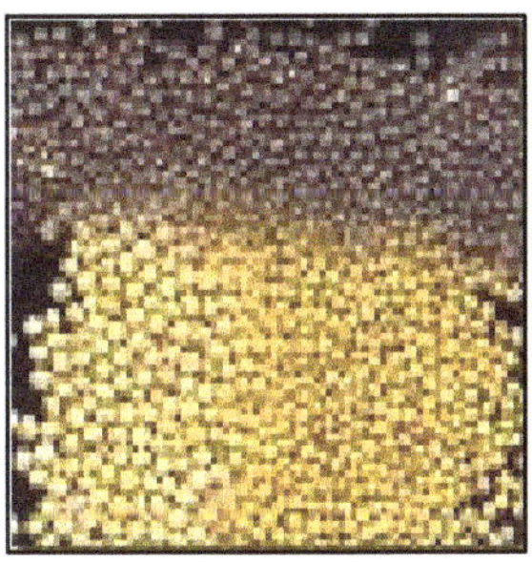

Figure 39

Figure 40

Figure 41

Figure 42

Figure 43

Figure 44

Figures 36–44: Various vegetables and fruits for prevention and cure of malnutritional cancer.

the development of cancer. Many doctors recommend that people wishing to reduce their risk of cancer eat several pieces of fruits and several portions of vegetables every day. An inverse relationship has been suggested between the consumption of fruits and vegetables, and the incidence of cancer in multiple organs, like lung, larynx, mouth, pharynx, gastrointestinal tract and pancreas[8,27–28,68]. The intake of 400-600 g per day of vegetables and fruits by a person can reduce the occurrence of many common forms of cancer, and diets rich in plant foods can also lower the risk of heart disease and many chronic diseases[8,28,69–70]. One-third of all cancer deaths in the United States could be avoided through dietary modification, which includes an abundant intake of fruits and vegetables. WHO study on diet, nutrition and prevention of chronic diseases recommended that we daily consume at least 400 g of vegetables and fruits including at least 30 g of pulses, nuts and seeds. People who eat much quantity of vegetables have about one-half the risk of cancer and less mortality from cancer[8–9,28].

Epidemiological data as well *in vitro* studies strongly suggest that vegetables and fruits have strong protective effect against major diseases including cancer. The protective action of fruits and vegetables against cancer has been attributed to the presence of antioxidants, especially antioxidant vitamins. Fruits and vegetables contain several phytochemicals which possess strong antioxidant activities. Thus the fruits and vegetables prevent from cancer and other diseases by protecting cells from damage caused by 'free radicals' highly reactive oxygen compounds. Certain phytochemical antioxidants with anticancer activity include vitamins (A, C, E, K), carotenoids, terpenoids, polyphenols (ellagic acid, gallic acid, tannins), flavonoids (quercetin, anthocyanins, catechins, flavones, flavonones, isoflavones), enzymes (superoxide dismutase, catalase, glutathion peroxidase), minerals (Cu, Mn, Se, Zn), polysaccharides, saponins, lignins and xanthones[2,5,7–9,11–12,20,24,27–28,32,71–73].

Vegetables and fruits contain compounds such as sulphoraphane that induces GSH transferase, there by helping detoxifying many types of carcinogens. Increased consumption of vegetables can increase the plasma antioxidant capacity and is associated with the lower risk of cancer. The vegetables and fruits are most effective against those cancers that involve epithelial cells such as cancers of lung, esophagus, stomach, colon, pancreas and cervix. The protective effect of vegetables has also been observed for hormone related cancers. The alcoholic extracts of dhaniya, haldi, karela and adrak have been tested for tumour growth inhibitory effects. Haldi and karela were most effective against metastatic prostate cancer cell lines. Vegetables with the highest anticancer activity included lasun, soybean, pattagobhi, gajar; with a modest level of cancer-protective activity include piyaz, nibu, haldi, phoolgobhi, tamatar; with a low level of anticancer activity included khira[8–9,20,27–28,68].

There are many fruits and vegetable pigments such as flavonoids, carotenoids and anthocyanins which protect us from various diseases. Quercetin (a flavonoid) chiefly present in red and yellow piyaz and French bean possesses both anticarcinogenic activity. The carotenoids are powerful antioxidants that provide protection against oxidative damage, and stimulate immune function. Persons with high levels of serum carotenoids have a reduced risk of cancer. In addition, a variety of phenolic compounds (caffeic, ellagic, and ferulic acids, sesamol and vanillin) are present in fruits and vegetables. These exhibit antioxidant and anticancer activities. Adrak contains phenolic compounds (gingerol) which have antioxidant activity that is even greater than α-tocopherol (vitamin E). Compounds that stimulate the activity of glutathione S-transferase (GST) are considered as inhibitors of cancer. Substances that stimulate GST activity include the phthalides in celery seeds; sulphides in piyaz and lasun; dithiolthiones and isothiocyanates in sarson, rai, phoolgobhi and pattagobhi; limonoids in nibu; and curcumins in adrak and haldi. The phytochemicals such as allyl sulphides present in lasun and piyaz; glucarates in nibu, began and alu; phytates, lignans, isoflavones and saponins in soybean; isothiocyanates in sarson, phoolgobhi and pattagobhi; and flavonoids, carotenoids and terpenoids in other fruits and vegetables are reported to provide protection against cancer. These block various hormone actions and metabolic pathways associated with cancer. The isoflavones present in soybean is the contributing factor in the low incidence of breast, prostate, stomach, colon, rectum and lung cancers. Soybean seed contains isoflavonoids which inhibit the growth of hormone-dependent and hormone-independent cancer cells in culture. The isoflavones in soy inhibit the growth of human breast and prostate cancers[2,8–9,28,69–70].

Nibu fruit due to its flavonoid, tangeretin and nobiletin contents can potentially inhibit the tumor cell growth and can activate the detoxifying cytochrome P-450 enzyme system. Limonoids present in nibu inhibit tumor formation by stimulating the enzyme GST. The limonene (a terpenoid) content of nibu also possesses anticancer activity. Nibu has been used for inhibiting the breast cancer cell proliferation, delaying of mammary tumorigenesis and curing the metastasis and leukemia cancers. Vegetables and fruits also contain a variety of isoprenoid compounds that exhibit antitumour/anticancer activities. These compounds include tocotrienols (related to tocopherols) and terpenoids (*e.g.*, limonene, geranoil, menthol and carvone). Overall, these compounds increase tumour latency and decrease tumour multiplicity[2,8–9,28,68–70].

Treatment of Anaemia by Edible Plants

Iron (Fe) is one of the most important components for the treatment of anaemia. Due to iron deficiency in the blood, the anaemia can occur. The edible plant sources of iron are the green vegetables and fruits. Therefore, to fulfil the minimum requirement of iron, one must eat the sufficient quantity of green vegetables (especially leafy) and fruits.

While breastfeeding is nearly universal in India, less than half of children (48 per cent) are fed only breast milk for the first 6 months, as recommended. At age 6 to 8 months, only about half of babies are given the recommended complementary foods along with breast milk. Only 44 per cent of breastfeeding children aged 6 to 23 months are fed at least the minimum recommended number of times per day and only 36 per cent are given the food from at least 3 food groups, as recommended to ensure adequate diversity in their diet. Overall, 21 per cent of breastfeeding and non-breastfeeding children are fed as per the recommended infant and young child feeding practices, including receiving breast milk (or other milk or milk products) and eating a minimum number of times per day from a minimum number of food groups that are appropriate for their age. The Government of India recommends twice yearly vitamin A supplements for children aged 9 to 36 months[51].

The first treatments for pernicious anaemia were devised by George Whipple, who studied anaemia in dogs. He was studying the effects of blood loss on dogs. He set out to investigate the effects of diet on anaemia, and bled dogs until they became anaemic, feeding them special diets and noting the effects on their recovery. He found that dogs which ate large amounts of liver recovered quickly from anaemia due to blood loss. He went on to investigate the effects of liver consumption on pernicious anaemia, and in 1920 he published a paper on how eating liver had beneficial effects on sufferers of pernicious anaemia[74]. George Minot and William Murphy investigated which factor in the liver was responsible for the treatment of anaemia in dogs, and found that it was iron. They also discovered that a soluble factor in the liver juices was responsible for the effects on pernicious anaemia. The fact that consuming liver treated both conditions turned out to be a coincidence. These findings were published in 1926[75], and they shared the Nobel Prize for medicine with Whipple in 1934 for the treatment for pernicious anaemia. The pernicious anaemia is no longer the fatal disease that it once was, and can be easily treated by injections of vitamin B_{12}[52].

The treatment of anaemia involves managing the underlying cause and, in some cases, blood transfusion. In some countries, there are established blood transfusion services for animals just as there are for humans[50]. The best way to prevent hypophosphatemia from "refeeding syndrome" is to reintroduce food gradually to malnourished patients, beginning with 25 per cent to 30 per cent of their resting energy requirement and gradually increasing to the full caloric requirement over 3 to 4 days[61].

Some Dietary Plants for Malnutritional Cancer in Humans and Animals

Use of plants for treating various diseases of both humans and animals is old practice. The dietary agricultural plants not only maintain the health and vitality of individuals, but also cure various diseases including cancer without causing any toxicity.

Some agricultural plants and their products such as vegetables, fruits and crops play an important role for the prevention of cancer. Many naturally occurring substances present in the diet/food have been identified as potential chemopreventive agents. The curative effect of primary stage of cancer is enhanced by combining the dietary supplementation (by agricultural foods) of metabolic deficiency with a simultaneous specific stimulation of the immunity of patients. This combined biological, agricultural and immunological therapy has cured hundreds of cancer patients and experimental animals in randomized studies over 30 years. These plants may promote host resistance against infection by re-stabilizing body equilibrium and conditioning the body tissues[20,25,27,31]. Many plant-derived products exhibit potent anticancer activity against several rodent and human cancer cell lines. A recently published study of 2400 Greek women noted that consumption of vegetables and fruits were independently associated with significant reductions in the incidence of breast cancer. Compared with meat eaters, most, but not all, studies have found that vegetarians are less likely to be diagnosed with cancer. Vegetarians have also been shown to have stronger immune function, possibly explaining why vegetarians may be partially protected against cancer[27,76].

There are about 400 families of the flowering plants in the world, at least 315 are represented by India[77]. Considerable works have been done on ethnomedicinal plants of India. But very little scientific attention has been paid to plants and other natural products, which are specifically used in the treatment of livestock diseases[78]. Various agriculture plants and their products have been used in veterinary practices since times immemorial. The dependence of rural mass on the plant-based medicines for curing animals is mainly because of the limited access to the modern (allopathic) medicine system, cheaper and easy availability, and the simplicity of their applications. The herbal drugs also do not possess toxicity or have very less side effects[79]. Ethnomedicinal survey exhibits a number of valuable data on indigenous medicinal plants, which are cultivated for agriculture purpose. Many agriculture plants and their products have been used traditionally for the treatment of various livestock diseases. These plants are used not only for food but also employed to cure different ailments of livestock. The medicinal uses of these plants may help veterinarians, medicos and agriculture/bio-scientists or researchers in discovering new drugs against various diseases of human and animals[20,80].

Many dietary agricultural plants and their products have been reported to possess versatile anticancer and immunomodulatory properties. They are commonly consumed by humans as well as animals as food matters, and prevent from malnutritional cancer and various other diseases. Some of those dietary plants are: *Allium cepa, A. sativum, Brassica campestris, B. oleracea, B. rapa, Citrus limon, Curcuma longa, Emblica officinalis, Glycine javanica, Lycopersicon esculentum, Momordica charantia, Swertia chirata, Trigonella foenumgraecum, Triticum aestivum* and *Zingiber officinale*[20,27,80]. Apart from these, some other useful agriculture plants for livestock diseases are: *Annona squamosa, Brassica nigra, Cajanus cajan, Capsicum annum, Carica papaya, Cicer arietinum, Daucus sativa, Elephantopus scaber, Lens culinaris, Lycopersicon esculentum, Oryza sativa, Picrorhiza kurroa, Psidium guajava, Raphanus sativus, Sesamum orientale, Solanum melongena, Solanum tuberosum, Spinacia oleracea, Trachyspermum ammi* and *Zea mays*[20,80]. Besides above, some more dietary plants which have been reported to exhibit anticancer properties include *Amorphophallus companulatus* (Suran), *Avena sativa* (Oat), *Cajanus cajan* (Arhar), *Hordeum vulgare* (Jau), *Lens culinaris* (Masur), *Mentha arvensis* (Podina) and *Zea mays* (Makka)[27,81–83].

Here, many of the above edible plants have been discussed below with special reference to their usefulness against cancer and other malnutritional diseases in humans and animals both. In general, the main active components, curative uses and some important biological activities of these plants have been mentioned by many authors[27,77,80–87]. However, some other authors have also reported specifically as elucidated under these plants.

1. *Allium cepa* Linn. (Piyaz)

Its bulb contains diallyl disulphide (a sulphurous product), quercetin flavonoid, allicin, allin, amino acids and vitamins (folic acid, C and E). Diallyl disulphide inhibits cancerous cells in stomach. Quercetin, because of its antioxidant properties, may treat lung cancer and some other cancers[68]. In animals, the boiled bulbs are given in flatulence and indigestion[79].

2. *Allium cepa* var. *aggregatum* (Shallot)

Its bulb contains allyl propyl disulphide (a sulphurous product), flavonoid, phenolic content, proteins and vitamins (folic acid and C). Allyl propyl disulphide, phenolic content and flavonoid inhibit cancerous cells in liver, stomach and other organs. Shallots help liver to eliminate toxins to inhibit and kill cancer cells[88].

3. *Allium sativum* Linn. (Lasun or lahsun)

The bulb (lahsun) of *A. sativum* contains sulphur compounds (diallyl sulphide, diallyl disulphide, allyl propyl disulphide), allicin, allin and amino acids. The sulphur compounds inhibit cell proliferation, modulate cell cycle activity and interfere with hormone action in cancer cells. Allicin inhibits the proliferation of cancerous cells of human mammary gland, endometrium and colon. Lasun is also used in gastric and liver cancers[69,89]. In animals, the bulbs also are given in malnutritional disorders of skin and lung, flatulence, and dyspepsia.

4. *Annona squamosa* Linn. (Sitaphal)

This plant contains annoaine, higenamine and amino acids. Its leaves and seeds are externally applied over malnutritional wounds, boils and swelling in animals[79]. The riped fruit (sitaphal) is rich of carbohydrate (energy), so it can be very well given to cure marasmus PEM.

5. *Brassica campestris* Linn. (Sarson)

Its seed oil contains glycerides of palmitic, stearic, oleic and linoleic acids. It is used in various tumours. In animals, the oil is also given in rheumatism, cancer and ulcer[68,79].

6. *Brassica juncea* (*B. nigra*) Linn. (Rai)

Its plant contains essential oils, while its seed contains 2 antithiamines and flavonol glycosides. In animals, oil along with rock salt is used in the diseases of gum[79]. The seeds and oil of rai are used in malnutritional rheumatism, paralysis and skin diseases.

7. *Brassica oleracea* var. *botrytis* Linn. (Phoolgobhi)

Its sprouts (flowers) contain cysteine, ascorbigen, sulphoxide, indole-3-carbinol, glucaric acid, sulphoraphane, glucosinolates, isothiocyanates and vitamins (A and C). These components possess anticancer activities against lung, stomach, colon, bladder, mammary and rectum cancers[69,90]. Its leaves contain vitamin C, and are externally used in gout, rheumatism, skin diseases and blister in animals; while phoolgobhi is cardiotonic and antiinflammatory, and may be used in biliousness and urinary diseases.

8. *Brassica oleracea* var. *capitata* Linn. (Pattagobhi, cabbage)

It contains glucaric acid, sulphoraphane, glucosinolates, isothiocyanates, cysteine, ascorbigen, sulphoxide, indole-3-carbinol, allyl isothionate and vitamins (A, B and C). These have anticancer activities against bladder, lung, stomach, colon, rectum and mammary cancers[69,90].

9. *Brassica rapa* Linn. (Shaljam)

Its leaves and roots contain ascorbigen and vitamins (A and C), and are used in various tumours[68–69].

10. *Cajanus cajan* Linn. (Arhar)

Its leaves contain cajanol, while roots contain cajanone and cajaquinone. Arhar seeds are quite rich in proteins, so the seeds (or cooked 'dal') can be highly given in both marasmus and kwashiorkor PEM. The expressed juice of leaves, with salt is given in animals for the treatment of malnutritional jaundice.

11. *Capsicum annum* Linn. (Lalmirch)

Its fruits contain amino, ascorbic and folic acids, glutathione, and flavonoids. In animals, the fruit powder with oil cures foot diseases[79]. Fruits are stimulant and stomachic, and are used in dyspepsia, neuralgia, rheumatism and diarrhoea.

12. *Carica papaya* Linn. (Papita)

Its fruit contains papin and cryptoflavin, while the seeds contain carpasemine. In animals, the seeds and fruits are given to cure hepatitis, diarrhoea and dysentery. The fruits are digestive, stomachic, carminative and diuretic too.

13. *Cicer arietinum* Linn. (Chana)

Its seed contains isoliquiritigenin, while the plant contains amino acids and vitamins (A, D and E). Its fresh leaves juice is given to livestock in constipation[78–79]. The macerated seeds in water are used as tonic. The seeds are astringent, diuretic and antibilious, and are given in debility, scurvy and jaundice. Chana seeds (or cooked 'dal') are quite rich in proteins, so the seeds (or cooked 'dal') can be very useful in marasmus and kwashiorkor PEM diseases.

14. *Citrus limon* Linn. (Baranibu)

Its fruit contains flavonoid, flavone, limonoid, limonene, nobiletin, tangeretin and vitamin C. The flavonoid, tangeretin and nobiletin are potent inhibitors of tumor cell growth and can activate the detoxifying

P-450 enzyme system. Limonoids inhibit tumour formation by stimulating the enzyme glutathione S-transferase (GST). The limonene (a terpenoid) also possesses anticancer activity. Nibu fruit is used for inhibition of human breast cancer cell proliferation and delaying of mammary tumorigenesis. It is also used in metastasis and leukemia[69,91]. The Leaves of C. *limon* contain bergapten, citropten and limettin. The fruits are antiseptic, appetizing, astringent and stomachic, and are given in scurvy, rheumatism, dysentery, diarrhoea and liver disorders of animals.

15. *Curcuma longa* Linn. (Haldi)

The rhizome (called haldi) of *C. longa* contains curcumin, zingiberine and curcuminoids. Haldi is reported to act against colon, bladder and prostate cancers, intravesical tumour, fibrosarcoma, hepatocellular carcinoma, oesophagal carcinogenesis, leukaemia, stomach papilloma and solid tumour cell lines[92–94]. In large animals, the rhizome powder is used in pain, swelling, wound, mastitis, tumour and liver diseases.

16. *Daucus sativa* DC. (Gajar)

Its root contains pyruvate kinase, acyltransferase and alkane; while the seeds contain sitosterol and amino acids. The boiled roots are given to cattle to fatten them. The roots and seeds are stimulant and carminative, and are used in dropsy and kidney diseases.

17. *Elephantopus scaber* Linn. (Gobhi)

In animals, the decoction of roots and leaves are given in dysuria, diarrhoea, dysentery, swelling or pain in stomach, ulcer and eczema. The whole plant is astringent and cardiac tonic.

18. *Emblica officinalis* Gaertn. (Amla)

Its fruit (called amla) contains ascorbic acid (vitamin C) and phyllembic acid. It inhibits cellular mutation and may prevent cancer. Its 18 compounds inhibit the growth of gastric and uterine cancer cells. Amla has also inhibited the growth of *in vitro* human breast cancer cells[2,95–96].

19. *Glycine javanica* Linn. (Soybean)

Soybean seed contains phytates, protease inhibitors, phytosterols, saponins, isoflavonoids and isoflavones. The soybean (soyabean or soy) is contributing factor in the low incidence of breast, prostate, stomach, colon, rectum and lung cancers. Its isoflavonoids inhibit the growth of hormone-dependent and hormone-independent cancer cells in culture; however, its isoflavones inhibit the growth of human breast and prostate cancers[69]. The soy seeds (or cooked 'dal') are quite rich in proteins, so the seeds (vegetables) may potentially treat both marasmus and kwashiorkor PEM diseases.

20. *Lens culinaris* Medic. (Masur)

The plant contains lenticin, tricetin and luteolin; while seed contains protein. In large animals, masur gruel is given in haematuria, and paste of seeds is applied on the wounds and broken horns[78]. The seeds are laxative and given in constipation. Since the seeds of *L. culinaris* are rich in protein, these (as masur dal) can be used to cure both marasmus and kwashiorkor PEM diseases.

21. *Lycopersicon esculentum* Mill. (Tamatar)

Its fruits contain lycopene. The lycopene, an antioxidant inhibits the proliferation of cancer cells. Higher intake of tamatar or higher blood levels of lycopene is correlated with protection from cancer; the protective effect was strongest for cancers of prostate, lung and stomach but some protective effects also appeared for cancers of pancreas, colon, rectum, esophagus (throat), mouth, breast and cervix[68–69,97]. Tomato fruits also contain oxalic, citric and mallic acids, and vitamins (A and C); while its seeds and herbage contain solanine. The pulp and juice of tomato fruits are intestinal antiseptic, blood purifier, cholagogue, stimulant and digestive, and are given in pulmonary diseases and dyspepsia of animals.

22. *Momordica charantia* Linn. (Karela)

Leaf, fruit (karela) and seed of *M. charantia* contain linoleic acid, momordin, palmitic acid, proteolytic enzymes and vitamins (A, B and C). These parts are used in colon, mammary and bladder carcinomas, lymphoma, leukemia and other tumours[93]. The fruits and seeds are carminative and stomachic, and are prescribed in rheumatism, gout and liver diseases of animals.

23. *Oryza sativa* Linn. (Chaaval)

Silver skin of *O. sativa* grain contains oryzanin; while bran contains a glucoside nukain, which yields aglucone nukagenin. In animals, chaaval with dal is used as a perfect food in dysentery and can alone support life as it contains protein, carbohydrate, fat, vitamins and minerals. Thus, chaaval may cure both marasmus and kwashiorkor PEM diseases. Decoction of grains is used internally as antiinflammatory for digestive tract.

24. *Picrorhiza kurroa* R. ex B. (Kutki)

Its fruits/seeds contain kutkin and kurrin. In animals, the fruits are given in loss of appetite, dropsy, fever, and foot and mouth disease (FMD)[79]. The roots are bitter, stomachic and purgative.

25. *Psidium guajava* Linn. (Amrud)

The plant contains leucocyanidin and quercetin, while the leaves contain eugenol. In animals, the decoction of root bark and leaves is given in chronic diarrhoea. Its root bark is astringent and fruits are laxative; while the leaves are astringent, and applied in wounds and ulcers.

26. *Raphanus sativus* Linn. (Muli)

Its root, *i.e.* muli contains glucoside, enzymes and methyl mercaptan. In animals, the roots are used in dysentery, indigestion, urinary disorders and gastrointestinal pain[79]. Its seeds are diuretic, laxative and carminative.

27. *Sesamum orientale* Linn. (Til)

Its seeds contain essential oil, proteins sesamin and sesamolin, so its seeds and oils can be given in PEM diseases of animals.

28. *Solanum melongena* Linn. (Baingan)

Its fruit contains vitamins A, B and C; while seed contains linoleic and oleic acids. In animals, the fruits and leaves are given in weakness, dysuria, liver disorders and ulcers. The seeds of *S. melongena* are stimulant.

29. *Solanum tuberosum* Linn. (Alu)

Its plant contains solanine, solasodine and lutein. In livestock, a portion of raw tuber (alu) is grated over the sore to cure leg ulcers. Alu is full of carbohydrate, it can be given in weakness, particularly in PEM.

30. *Spinacia oleracea* Linn. (Palak)

Its plant contains vitamins, iodine, lecithin, carotene and amino acids. The whole plant and leaves of *S. oleracea* are used in inflammation of lung, liver and bowel, fever and jaundice in animals. Its seeds and fruits are laxative, demulcent and diuretic.

31. *Trachyspermum ammi* Linn. (Ajwain)

Its fruits yield essential oil containing thymol. In animals, the fruits are given in weakness after pregnancy and low lactation[98]. They are also employed in gastric disorders and diarrhoea, and have antiseptic, carminative and stimulant properties.

32. *Trigonella foenumgraecum* Linn. (Methi)

Its leaves and seeds contain choline, trigonelline, amino acids, proteins, vitamins and quercetin. These parts of methi are used in various cancers[92-93]. In animals, the leaves and seeds are used in rheumatism, cancer, indigestion and bilious disorders. Its seeds are carminative and tonic too.

33. *Triticum aestivum* Linn. (Gehun)

The grains of *T. aestivum* contain ellagic, linoleic and oleic acids, oligosaccharides, fat, sterols, phytases, tocotrienols and vitamin E. Whole grains protect against lymphomas and cancers of pancreas, stomach, colon, rectum, breast, uterus, mouth, throat, liver and thyroid[68]. In animals, the flour of wheat is given in weakness and gastric disorders.

34. *Zea mays* Linn. (Makka)

Its grains contain oxalic acid, carbohydrate, fat, protein and vitamins. The grains are astringent, nutritive and nourishing in humans and animals.

35. *Zingiber officinale* Linn. (Adrak or sonth)

The rhizome (called adrak and the dried rhizome called as sonth) contains a volatile oil containing camphene, gingerol, zingiberene, borneol, cineol and proteins. It is used in various cancers[92]. In livestock animals, the rhizomes are given with salt before feed to increase appetite and in weakness after pregnancy[79,98]. They are also given in swelling, indigestion, flatulent colic, and are stimulant and carminative.

6
References

1. Loose, D.S. and Stancel, G.M. (2006). Estrogens and progestins. In: *Goodman & Gilman's The Pharmacological Basis of Therapeutics,* 11th Edn. (Brunton, L.L., ed.). McGraw-Hill Co., New York. pp. 1541-1571.
2. Madhuri, S. (2008). *Studies on oestrogen induced uterine and ovarian carcinogenesis and effect of ProImmu in rat.* Ph.D. thesis. Rani Durgavati Vishwavidyalaya, Jabalpur, M.P., India.
3. Pandey, Govind (2007). *Role of malnutrition in causing cancer and its safety measures.* Diploma in Nutrition and Health Education (DNHE) project report. Indira Gandhi National Open University, New Delhi, India.
4. Pandey, Govind (2010). Malnutrition leading to cancer by some environmental hazards. Int. J. Res. Ayur. Pharm., 1(2): 287-291.
5. Pandey, Govind and Madhuri, S. (2009). Some medicinal plants as natural anticancer agents. PHCOG REV., 3(6): 259-263.
6. Somkuwar, A.P. (2003). *Studies on anticancer effects of Ocimum sanctum and Withania somnifera on experimentally induced cancer in mice.* Ph.D. thesis. J.N.K.V.V., Jabalpur, M.P., India.
7. Kathiresan, K., Boopathy, N.S. and Kavitha, S. (2006). Coastal vegetation- An underexplored source of anticancer drugs. Nat. Prod. Rad., 5(2): 115-119.

8. Pandey, Govind and Madhuri, S. (2010). Significance of fruits and vegetables in malnutrition cancer. Pl. Arch., 10(2): 517-522.

9. Pandey, Govind and Madhuri, S. (2011). Therapeutic approach to cancer by vegetables with antioxidant activity. Int. Res. J. Pharm., 2(1): 10-13.

10. Kumar, V., Abbas, A.K. and Fausto, N. (2006). In: *Pathologic Basis of Disease*, 7th Edn. Saunders, Elsevier India Pvt. Ltd., New Delhi, India. pp. 269-342 and 1059-1117.

11. Pandey, Govind (2009). An overview on certain anticancer natural products. J. Pharm. Res., 2(12): 1799-1803.

12. Pandey, Govind and Madhuri, S. (2008). Some anticancer agents from plant origin. Pl. Arch., 8(2): 527-532.

13. Reddya, L., Odhava, B. and Bhoolab, K.D. (2003). Natural products for cancer prevention: A global perspective. Pharmacol. and Therap., 99: 1-13.

14. DNHE-3 (2004). Nutrition and health status of the community. In: *Nutrition and Health Education*. Indira Gandhi National Open University, New Delhi, India.

15. Greenberg, D.S. (1978). Washington report: The Cancer Institute on the Griddle. *N. Engl. J. Med.*, 299(4): 207-208.

16. Pupim, L.B., Kent, P. and Hakim, R. (1999). The potential of intradialytic perenteral nutrition: A review. Minor Electrolyte Metab., 25: 317-323.

17. DNHE-1 (2004). Nutrition-related disorders. In: *Nutrition for the Community*. Indira Gandhi National Open University, New Delhi, India.

18. Gottschlich, M.M. (2001). *The Science and Practice of Nutrition Support*. Kendall/Hunt Publishing, Dubuque.

19. DNHE-2 (2003). Dietary management of disease. In: *Public Health and Hygiene*. Indira Gandhi National Open University, New Delhi, India, 40.

20. Pandey, Govind and Sahni, Y.P. (2011). Phytotherapy of malnutritional cancers in animals. Int. J. Pharmaceu. Sci. Rev. Res., 8(1): 14-20.

21. Parkin, D.M., Bray, F., Ferlay, J. and Pisani, P. (2005). Global cancer statistics 2002. Cancer J. Clin., 55: 74-108.

22. Madhuri, S. and Pandey, Govind (2009). Some anticancer medicinal plants of foreign origin. Curr. Sci., 96(6): 779-783.

23. Pandey, Govind P. (1990). *Hepatogenic effect of some indigenous drugs on experimental liver damage.* Ph.D. thesis. JNKVV, Jabalpur, M.P., India.

24. Madhuri, S. and Pandey, Govind (2009). Ethnomedicinal plants for prevention and treatment of tumours. Int. J. Green Pharm., 3(1): 2-5.

25. Pandey, Govind and Madhuri, S. (2006). Medicinal plants: Better remedy for neoplasms. Indian Drugs, 43(11): 869-874.

26. Rosangkima, G. and Prasad, S.B. (2004). Antitumour activity of some plants from Meghalaya and Mizoram against murine ascites Dolton's lymphoma. Indian J. Exp. Biol., 42: 981-988.

27. Madhuri, S. and Pandey, Govind (2008). Some dietary agricultural plants with anticancer properties. Pl. Arch., 8(1): 13-16.

28. Pandey, Govind (2011). Antioxidant vegetables act against cancer and other diseases. Int. J. Pharmaceu. Stud. Res., 1(2): 32-38.

29. Sivalokanathan, S., Ilayaraja, M. and Balasubramanium, M.P. (2005). Efficacy of *Terminalia arjuna* (Roxb.), an N-nitrosodiethylamine induced hepatocellular carcinoma in rats. Indian J. Exp. Biol., 43: 264-267.

30. Agrawala, S.K., Chatterjee, S. and Misra, S.K. (2001). Immune-potentiation activity of a polyherbal formulation "Immu-21" (Research Name). Phytomedica, 2(1 and 2): 1-22.

31. Pandey, Govind and Madhuri, S. (2006). Autochthonous herbal products in the treatment of cancer. Phytomedica, 7: 99-104.

32. Polidori, M.C. (2003). Antioxidant micronutrients in the prevention of age-related diseases. J. Postgrad. Med., 49: 229-235.

33. Copeland, E.M. 3rd, MacFadyen, B.V. Jr., Lanzotti, V.J. and Dudrick, S.J. (1975). Intravenous hyperalimentation as an adjunct to cancer chemotherapy. *Am. J. Surg.*, 129(2): 167-173.

34. DNHE-3 (2004). Nutrition programmes. In: *Nutrition and Health Education*. Indira Gandhi National Open University, New Delhi, India.

35. Stock photos of malnutrition.htm (2011). Website.

36. Protein Energy Malnutrition.htm (2011). Website.

37. Merritt, R.J. (1998). *The A.S.P.E.N. Nutrition Support Practice Manual*. A.S.P.E.N. Silver Spring.

38. Photo of Cancer- RightHealth.htm (2011). Website.

39. Malnutrition Cartoons and Comics.htm (2011). Website.

40. Mehrotra, R. and Kopple, J.D. (2004). Causes of protein-energy malnutrition in renal failure. In: *Nutritional Management of Renal Disease*, 2nd Edn. (Kopple, J.D. and Massary, S.G., eds.). Lippincott Williams and Wilkins, Philadelphia. pp. 167-182.

41. Daugirdas, J.T. and Van Stone, J.C. (2001). Chronic hemodialysis prescription: A urea kinetic approach. In: *A Handbook of Dialysis*, 3rd Edn. (Daugirdas, J.T., Blake, P.G., Ing, T.S., eds.). Lippincott Williams and Wilkins, Philadelphia. pp. 121-147.

42. Pandey, Govind and Madhuri, S. (2010). Cancer pathogenesis caused by xenoestrogens of environment and food contaminants: A review. J. Chem. Pharm. Res., 2(4): 687-695.

43. Mukherjee, S., Bidhan, C.K., Ray, S. and Ray, A. (2006). Environmental contaminants in pathogenesis of breast cancer. Indian J. Exp. Biol., 44: 597-617.

44. Sandhu, H.S. (2006). Pesticides as endocrine disruptors (Abs.). Souvenir and Abstracts: 11-12 (International Conference and 26th Annual Meeting of Society of Toxicology, India, Gwalior, M.P., India, 9th-11th October, 2006).

45. Gold, L.S., Stone, T.H. and Ames, B.N. (1997). Prioritization of possible carcinogenic hazards in food. In: *Food Chemical Risk Analysis*. Chapman and Hall, New York. pp. 267-295.

46. Ray, S. (2001). *Studies on paraquat toxicity in rat*. Ph.D. thesis. University of Calcutta, Kolkata, W.B., India.

47. Sen, R. (2008). Oral cancer- The global scenario. Proceedings (Part II, Abstracts-Medical Sciences): 1-26 (95th Indian Science Congress of the

Indian Science Congress Associaton, Visakhapatnam, A.P., India, 3rd-7th January, 2008).

48. www.cat-world.com.au/anemia-in-cats (2012).
49. http://fabcats.org/owners/anaemia/info1.html (2012).
50. www.provet.co.uk/petfacts/healthtips/anaemia.htm (2012).
51. http://hetv.org/india/nfhs/nfhs3/NFHS-3-Nutrition-and-Anaemia.pdf (2012).
52. www.animalresearch.info/en/medical-advances/158/pernicious-anaemia (2012).
53. Borelli, P., Blatt, S., Pereira, J., de Maurino, B.B., Tsujita, M., de Souza, A.C., Xavier, J.G. and Fock, R.A. (2007). Reduction of erythroid progenitors in protein-energy malnutrition. Br. J. Nutr., 97(2): 307-314.
54. www.petmd.com/horse/conditions/cardiovascular/c_hr_anemia#.UJBOsaCNBJE (2012).
55. www.ifahsec.org/animal-health/parasites (2012).
56. www.largeanimalhospital.com/articles/27-equine-infectious-anemia.html (2012). Davie County Large Animal Hospital, Farmington Road, Mocksville, NC.
57. www.2ndchance.info/anemia.htm (2012).
58. www.petplace.com/dogs/anemia-in-dogs/page1.aspx (2012).
59. www.ehow.com/about_5061451_signs-symptoms-dogs-anemia.html (2012).
60. http://pets.webmd.com/dogs/symptoms-and-treatments-of-anemia-in-dogs (2012).
61. www.alvedia.com/pdf/publications/ANEMIA_IN_FELINE_CRITICAL_ILLNESS (2012).
62. www.cat-world.com.au/anemia-in-cats (2012).
63. The Gentle Barn, Santa Clarita, CA (2012). Website.
64. Animal Specialty Hospital, Rockledge Florida.htm. (2011). Website.
65. Natural-Dog-Health-Remedies.com (2011). Website.

66. Animal World-Brook Hospital for Animals.htm (2011). Website.

67. Shopbell, J.M., Hopkins, B. Shronts, E.P. (2001). Nutrition screening and assessment. In: *The Science and Practice of Nutrition Support* (Gottschlich, M.M., ed.). Kendall/Hunt Publishing, Dubuque.

68. Vecchia, C.L. and Tawani, A. (1998). Fruits, vegetables and human cancer. Eur. J. Cancer, 7: 3-8.

69. Heber, D. (2004). Vegetables, fruits and phytoestrogens in the prevention of diseases. J. Postgrad. Med., 50: 145-149.

70. Craig WJ. Phytochemicals: Guardians of our health website. 2006.

71. Gupta,V.K. and Sharma, S.K. (2006). Plants as natural antioxidants. Nat. Prod. Rad., 5(4): 326-334.

72. Madhuri, S. and Pandey, Govind (2010). Effect of ProImmu, a herbal drug on estrogen caused uterine and ovarian cytotoxicity. biomed, 5(1): 57-62.

73. Ray, G. and Hussan, S.A. (2002). Oxidant, antioxidant and carcinogenesis. Indian J. Exp. Biol., 40: 1213-1232.

74. Whipple, G.H., Robscheit, F.S. and Hooper, C.W. (1920) Blood regeneration following simple anemia. IV. Influence of meat, liver and various extractives, alone or combined with standard diets. Am. J. Physiol., 53: 236-262.

75. Minot G.R. and Murphy, W.P. (1926). Treatment of pernicious anemia by a special diet, JAMA, 87: 470-476.

76. Armstrong, B.K., Brown, J.B. and Clarke, H.T. (1981). Diet and reproductive hormones: a study of vegetarian and nonvegetarian postmenopausal women. *J. Natl. Cancer Inst.*, 67: 761-767.

77. Sharma, R. (2003). *Medicinal Plants of India- An Encyclopedia.* Daya Publishing Home, Delhi, India.

78. Tiwari, L. and Pande, P.C. (2006). Indigenous veterinary practices of Drama valley of Pithoragarh district, Uttarakhand. Indian J. Trad. Knowledge, 5(2): 201-206.

79. Jadeja, B.A., Odedra, N.K., Solanki, K.M. and Baraiya, N.M. (2006). Indigenous animal healthcare practices in district Porbandar, Gujarat. Indian J. Trad. Knowledge, 5(2): 253-258.

80. Pandey, Govind and Madhuri, S. (2008). Agriculture plants used against livestock diseases. Int. J. Plant Sci., 3(2): 671-674.

81. C.S.I.R. (1986). The Useful Plants of India. New Delhi, India.

82. Kaushik, P. and Dhiman, A.K. (1999). *Medicinal Plants and Raw Drugs of India.* Bishen Singh Mahendra Pal Singh, Dehra Dun, Uttarakhand, India.

83. Prajapati, N.D., Purohit, S.S., Sharma, A.K. and Kumar, T. (2003). *A Hand Book of Medicinal Plants,* 1st Edn. Agrobios (India).

84. Chopra, R.N., Nayar, S.L. and Chopra, I.C. (2002). *Glossary of Indian Medicinal Plants.* Publications and Information Directorate, C.S.I.R., New Delhi, India.

85. I.D.M.A. (2002). *Indian Herbal Pharmacopoeia.* Mumbai, India.

86. Jarald, E.E. and Jarald, S.E. (2006). *Colour Atlas of Medicinal Plants.* New Delhi.

87. Nadkarni, K.M. (1954). *Indian Materia Medica* (Vol. 2), 3rd Edn. Popular Book Depot, Bombay and Dhootpapeshwar Prakashan Ltd., Bombay (Mumbai), India.

88. Cherry, A. (2006). Alternative medicine in food-shallots. Meta Suchmaschine-seek XL de. Website.

89. Arivazhagan, S., Velmurugan, B., Bhuvaneswari, V. and Nagini, S. (2004). Effect of aqueous extract of garlic (*Allium sativum*) and neem (*Azadirachta indica*) leaf containing blood oxidant-antioxidant status during experimental gastric carcinogesis. J. Med. Food., 7(3): 334-339.

90. Beecher, C.W.W. (1994). Cancer preventive properties of varieties of *Brassica oleracea*: A review. *Am. J. Clin. Nutr.,* 59 (suppl): 1166-1170.

91. Hirano, T., K. Oka, Y. Mimaki, M. Kuroda and Y. Sashida (1996). Potent growth inhibitory activity of a novel ornithogalum cholestane glycoside on human cells: induction of apoptosis in promyelocytic leukemia HL-60 cells. Cancer Lett., 58: 789-798.

92. Aruna, K. and Sivaramkrishanan, V.M. (1990). Plant products as protective agents against cancer. Indian J. Exp. Biol., 28: 1008-1011.

93. Rao, K.V.K., Schwartz, S.A., Nair, H.K., Aalinkeel, R., Mahajan, S., Chawda, R. and Nair, M.P.N. (2004). Plant derived product as a

source of cellular growth inhibitory phytochemical on PC-3M, DU-145 and LNCaP prostate cancer cell lines. Current Science, 87(11): 1585-1588.

94. Singhal, S.S., Awasthi, S. and Pandya, U. (1999). The Effect of curcumin on glutathione linked enzymes in K562 human leuklemia cells. Toxicol. Lett., 109: 87-95.

95. Lambertini, E., Piva, R., Khan, M.T., Lampronti, I., Bianchi, N., Borgatti, M. and Gambari, R. (2004). Effect of Bangladeshi medicinal plants on *in vitro* proliferation of human breast cancer cell lines and expression of estrogen receptor alpha gene. Int. J. Oncol., 24(2): 419-423.

96. Zhang, Y.J., Nagao, T., Tanaka, T., Yang, C.R., Okabe, H. and Kouno, I. (2004). Antiproliferative activity of the main constituents from *Phyllanthus emblica*. Biol. Pharm. Bull., 27(2): 251-255.

97. Giovannucci, E. (1999). Tomatoes, tomato-based products, lycopene, and cancer: Review of the epidemiologic literature. *J. Natl. Cancer Inst.*, 91: 317-331.

98. Jadhav, D. (2006). Ethnomedicinal plants used by Bhil tribe of Bibdod, Madhya Pradesh. Indian J. Trad. Knowledge, 5(2): 263-267.

7

List of Published Papers and Project Report Related to this Book

I. Papers Published

1. Sharma, Madhuri, Pandey, Govind P. and Gupta, V.B. (2007). Common agriculture plants as anticancer (Abs.). Abstracts (I): 314-315 *(International Conference on Sustainable Agriculture for Food, Bio-energy and Livelihood Security*, February 14-16, 2007, JNKVV, Jabalpur).
2. Sharma, Madhuri, Pandey, Govind and Khanna, A. (2007). Antioxidant vegetables useful in cancers and other diseases (Abs.). Abstract: 341-342 *(1st Bhartiya Vigyan* Sammelan, November 23-25, 2007, Madhya Pradesh Council of Science & Technology, Bhopal, MP).
3. Pandey, Govind P., Sharma, Madhuri and Gupta, V.B. (2007). Some agriculture plants used in livestock diseases (Abs.). Abstracts (I): 248 *(International Conference on Sustainable Agriculture for Food, Bio-energy and Livelihood Security*, February 14-16, 2007, JNKVV, Jabalpur, MP).
4. Sharma, Madhuri and Pandey, Govind (2008). Some dietary agricultural plants with anticancer properties. Pl. Arch., 8(1): 13-16 *[Indexed in Indian Science Abstracts,* 44 (14), 2008-013424; Medicinal and Aromatic Plants Abstr., 30(6): 875, 2008-06-3380].

5. Pandey, Govind and Sharma, Madhuri (2008). Agriculture plants used against livestock diseases. *Int. J. Plant Sci.,* 3(2): 671-674.

6. Sharma, Madhuri and Pandey, Govind (2009). Environmental contaminants causing malnutrition and cancer. (Abs.). Abstract: 45 *(National Seminar on Environmental Hazards and Disasters*, January 12-13, 2009, Hawabagh Women's College, Jabalpur).

7. Pandey, Govind (2009). Cancer pathogenesis due to xenoestrogens of environments and food contaminations (Abs.). Abstracts: 47 *(International Symposium on Current* Trends in Toxicology and Pre-clinical Toxicology *and XXIX Annual Conference of Society of Toxicology,* India, November 5-7, 2009, National Institute of Nutrition, Hyderabad, AP).

8. Pandey, Govind and Sharma, Madhuri (2009). आधुनिक जीवन शैली एवं कुपोषण से कैंसर: एक कयामत (Abs.). स्मारिका एवं सारांश: 92 *(Madhya Kshetriya Vigyan Sammelan, Feb.* 21-22, 2009, Govt. MH College of Home Science and Science for Women, Jabalpur).

9. Pandey, Govind and Sharma, Madhuri (2009). Radiations in the pathogenesis of cancer. Jigyasa, 3: 86-90.

10. Pandey, Govind, Sharma, Madhuri and Khanna, A. (2009). Radiations as environmental hazard and disaster in cancer pathogenesis (Abs.). Abstract: 21 *(National Seminar on Environmental Hazards and Disasters*, January 12-13, 2009, Hawabagh Women's College, Jabalpur).

11. Pandey, Govind (2010). Malnutrition leading to cancer by some environmental hazards. Int. J. Res. Ayur. Pharm., 1(2): 287-291.

12. Pandey, Govind and Sharma, Madhuri (2010). Cancer pathogenesis caused by xenoestrogens of environment and food contaminants: A review. J. Chem. Pharm. Res., 2(4): 687-695.

13. Pandey, Govind and Sharma, Madhuri (2010). Significance of fruits and vegetables in malnutrition cancer. Pl. Arch., 10(2): 517-522.

14. Pandey, Govind and Sharma, Madhuri (2010). आधुनिक जीवन शैली एवं कुपोषण से कैंसर: एक कयामत Science India (Hindi), 3(3): 24-26.

15. Pandey, Govind, Pandey, S.P. and Sharma, Madhuri (2010). Significance of fruits and vegetables in malnutrition cancer (Abs.).

Souvenir: 21 *(National Conference on Biodiversity Conservation: Problems and Prospects,* October 6, 2010, Hawabagh Women's College, Jabalpur).

16. Pandey, Govind and Sharma, Madhuri (2010). Global scenario of radiation in cancer risk: A review. Int. J. Pharmac. Biol. Arch., 1(3): 195-198.
17. Pandey, Govind and Sharma, Madhuri (2010). Some plants as source of anticancer drugs (Abs.). Souvenir: 38-39 *(National Seminar on Current Advances in Bioscience: Application in Health, Environment and Agriculture,* November 23-24, 2010, Hitkarini Mahila Mahavidyalaya, Jabalpur).
18. Pandey, Govind (2011). Antioxidant vegetables act against cancer and other diseases. Int. J. Pharmaceu. Stu. Res., 2(1): 32-38.
19. Pandey, Govind (2011). Some important anticancer herbs: A review. Int. Res. J. Pharm., 2(7): 45-52.
20. Pandey, Govind and Sharma, Madhuri (2011). Therapeutic approach to cancer by vegetables with antioxidant activity. Int. Res. J. Pharm., 2(1): 10-13.
21. Pandey, Govind and Sahni, Y.P. (2011). Phytotherapy of malnutritional cancers in animals. Int. J. Pharmaceu. Sci. Rev. Res., 8(1): 14-20.

II. Project Report Submitted and Approved

1. Pandey, Govind (2007). Role of malnutrition in causing cancer and its safety measures. Project Report of Diploma in Nutrition and Health Education (DNHE), submitted to Indira Gandhi National Open University, New Delhi.

8

Text of Published Papers Related to this Book

Some Agriculture Plants Used in Livestock Diseases

Govind P. Pandey**, Madhuri Sharma and V.B. Gupta*

**B.R. Nahata College of Pharmacy and B.R.N.S.S. Contract Research Center, Mandsaur & College of Veterinary Science and A.H., J.N.K.V.V., Jabalpur- 482001 (M.P.), India*

Various agriculture plants and their products have been used in veterinary practices since times immemorial. However, a very little scientific attention has been paid on this field. Ethnomedicinal survey exhibits a number of valuable data on indigenous medicinal plants, which are cultivated for agriculture purpose. Thus, the present paper signifies the efficacy of some agriculture plants being used against various diseases of livestock. Cattle, buffalo, sheep and goat are the main livestock and domestic animals. Some important agriculture plants traditionally used against different diseases of livestock are: *Raphanus sativus* (Muli) in dysentery and indigestion; *Picrorhiza kurroa* (Kutki) in loss of appetite, dropsy, fever, foot and mouth disease (FMD) and internal parasite; *Lens culinaris* (Masur) in haematuria, constipation, intestinal affections and broken horns; *Zingiber officinale* (Adrak) in swelling, vomiting, indigestion, flatulent colic, diabetes and heart diseases; *Curcuma longa* (Haldi) in pain, sprain, swelling, cut, wound, cough, cold, liver diseases, tumour, loss of

---

[Abstract paper published in: Abstracts (I) of International Conference of JNKVV, Jabalpur, MP held on February 14-16, 2007; p. 248].

appetite and weakness; *Triticum aestivum* (Gehun) in food poisoning, low lactation, loss of appetite, gastric problem and general disorders of health; *Capsicum annum* (Lalmirch) in neuralgia, rheumatism, sore throat, dyspepsia, yellow fever, diarrhoea, snake bite, internal parasites, cough and cold; *Brassica campestris* (Sarson) and *B. nigra* (Rai) in scabies, external parasites, rheumatism, cancer, tumour, chronic cough, bronchial catarrh, dengue fever, stiff neck, skin eruption and ulcer; *Trachyspermum ammi* (Ajwain) in weakness after pregnancy, low lactation, colic, dyspepsia, diarrhoea, cholera, sore throat, bronchitis, pneumonia and stomach disorder; *Momordica charantia* (Karela) in rheumatism, gout, liver and spleen diseases, haemorrhoids and weakness; *Anisophyllea squamosa* (Sitaphal) in external parasite, boils, burn, swelling, wound and weakness; *Trigonella foenumgraecum* (Methi) in rheumatism, cancer, weakness, intestinal inflammation and low lactation; *Cicer arietinum* (Chana) in scurvy, deficiency diseases, jaundice and constipation; *Citrus limon* (Baranibu) in scurvy, rheumatism, dysentery and diarrhoea; *Lycopersicon esculentum* (Tamatar) in weakness and scurvy; *Solanum tuberosum* (Alu) in weakness, low lactation and cough; *Elephantopus scaber* (Gobhi) in dysuria, swelling, stomach pain, dysentery, diarrhoea, eczema and ulcer; *Allium cepa* (Piyaz) in general debility, flatulence, dysentery, swelling, pain, diabetes, cancer, heart disease, fever, cough, cold, blisters and boils; *Allium sativum* (Lasun) in pulmonary diseases, flatulence, colic, atonic dyspepsia and skin troubles; *Daucus sativa* (Gajar) in dropsy and kidney troubles; *Solanum melongena* (Baingan) in asthma, weakness, dysuria, bronchitis, liver disorder and ulcer; *Spinacia oleracea* (Palak) in inflammation of lung, liver and bowel, difficult breathing and urinary calculi.

Common Agriculture Plants as Anticancer

Madhuri Sharma[1], Govind P. Pandey[2] and V.B. Gupta[2]

[1]*College of Veterinary Science and A.H., J.N.K.V.V., Jabalpur and*
[2]*B.R. Nahata College of Pharmacy and*
B.R.N.S.S. Contract Research Center, Mandsaur (M.P.), India

Use of plants for treating various diseases of both animal and man is old practice. According to WHO estimates, more than 80 per cent people in developing countries depend on traditional medicine for their primary health needs. Since the allopathic system of medicine may cause several untoward effects on the normal host tissues, the indigenous system of medicine (medicinal plants) has become popular throughout the world. India is known to harbour a rich wealth of agriculture-medicinal plants, which provide high quality food and raw materials including drugs. Our country is the largest producer of plants and is rightly called the "Botanical Garden of the world". In India, 45,000 plant species have been identified, out of which about 15-20 thousand plants are of good medicinal value. However, traditional communities use only about 7000-7500 plants for medicinal purpose. The Siddha system of medicine uses about 600, Ayurveds 700, Unani 700 and modern medicine about 30 medicinal plants for treating variety of diseases in man and animal. However, only few medicinal plants have attracted the interest of scientists to investigate the

[Abstract paper published in: Abstracts (I) of International Conference of JNKVV, Jabalpur, MP held on February 14-16, 2007; pp. 314-315].

remedy for cancer. The agriculture plants not only maintain the health and vitality of individuals but may cure various diseases including cancer without causing any toxicity. A recent survey shows that more than 60 per cent of cancer patients use medicinal plants at some point in their therapy. The curative effect of primary stage of the cancer is enhanced by combining the dietary supplementation (by agriculture plants and their products) of the metabolic deficiency with a simultaneous specific stimulation of the immunity of the patients. This combined biological, agricultural and immunological therapy has cured hundreds of cancer patients and experimental animals in randomized studies over 30 years. The anticancer plants are known to have good immunomodulatory properties. These act by stimulating both non-specific and specific immunity. These plants may promote host resistance against infection by re-stabilizing body equilibrium and conditioning the body tissues. In view of this fact, the present paper aims to create awareness about some anticancer-agriculture plants and their products, which are used in our day-to-day life and also for livelihood of animals. Some common agriculture-medicinal plants which have been reported to possess anticancer activity against various types of malignant (cancer) and benign tumours are: *Allium cepa* (Piyaz), *Allium sativum* (Lasun), *Brassica campestris* (Sarson), *Brassica rapa* (Shaljam), *Brassica oleracea* (Koli), *Curcuma longa* (Haldi), *Emblica officinalis* (Amla), *Momordica charantia* (Karela), *Ocimum sanctum* (Tulsi), *Ricinus communis* (Arand), *Solanum nigrum* (Makoi), *Swertia chirata* (Chirayita), *Trichosanthes dioica* (Parwal), *Trigonella foenumgraecum* (Methi), *Triticum aestivum* (Gehun), *Withania somnifera* (Ashwagandha) and *Zingiber officinale* (Adrak).

Antioxidant Vegetables Useful in Cancers and Other Diseases

S. Madhuri[1], Govind Pandey[2] and Asha Khanna[3]

[1]Lecturer, Shri Ram Institute of Technology Pharmacy, Jabalpur
[2]Professor/Principal Scientist and Head, Department of Pharmacology, B.R. Nahata College of Pharmacy and Contract Research Center, Mandsaur
[3]Professor and Head, Department of Zoology and Biotechnology, Model Science College, Jabalpur (M.P.)

Cancer is a very severe disease, which kills annually about 3500 per million population allover the world. Several chemopreventive agents are used to treat cancers, but they cause toxicity that prevents their usage. Although more than 1500 anticancer drugs are in active development with over 500 of the drugs under clinical trials, there is an urgency to investigate much effective drugs with very low or no toxicity. Plants and their products have been used for treating various diseases of humans as well as animals since times immemorial. According to WHO estimates, more than 80 per cent people in developing countries depend on traditional medicine for their primary health needs. Ethnomedicinal survey exhibits a number of valuable data on indigenous medicinal plants, which cure different diseases including cancer without causing any toxicity. A recent survey shows that more than 60 per cent of patients use medicinal plants at some point in their therapy. As one of the major natural resources, the

---

(Abstract paper published in: Abstract of 1st Bhartiya Vigyan Sammelan organized by Madhya Pradesh Council of Science and Technology, Bhopal, MP held on November 23-25, 2007; pp. 341-342).

vegetables prevent human from several severe and chronic diseases. Many doctors recommend that people wishing to reduce their risk of cancer eat several pieces of fruits and several portions of vegetables every day. A report of the WHO study on diet, nutrition and prevention of chronic diseases recommended that we daily consume at least 400 g of vegetables and fruits including at least 30 g of pulses, nuts and seeds. People who eat much quantity of vegetables have about one-half the risk of cancer and less mortality from cancer. The intake of 400-600 g per day of vegetables and fruits can reduce the occurrence of many common forms of cancers, heart disease and diseases. Thus, a balanced and regular diet with at least five portions of vegetables and fruits per day is required for a healthy life. The antioxidant vegetables prevent the human from many severe and chronic diseases including cancer, cardiovascular disease (CVD), diabetes, hypertension, leprosy, rheumatism, epilepsy, liver and urinary disorders, stroke, inflammation, paralysis, asthma, diarrhea, etc.

Vegetables contain several phytochemicals possessing antioxidant activity. Antioxidants protect the cells from damage caused by 'free radicals'- the highly reactive oxygen compounds. The major groups of the vegetable phytochemicals which shows antioxidant activity include vitamins (A, C, E, K), carotenoids, terpenoids, flavonoids, polyphenols, saponins, enzymes and minerals. All these compounds prevent from various cancers and other diseases. The examples of these phytochemicals which provide protection against cancer are: allyl sulphides present in lasun and piyaz; glucarates in nibu, began and alu; phytates, lignans, isoflavones and saponins in soybean; isothiocyanates in sarson, phoolgobhi and pattagobhi; and flavonoids, carotenoids and terpenoids in different vegetables. These phytochemicals block various hormone actions and metabolic pathways which are associated with the development of cancers. A large body of scientific evidence indicates that an association exists between inadequate antioxidant status and increased risk for many age-related diseases including cancer, congestive heart failure (CHF), CVD, peripheral arterial disease, osteoporosis, osteoarthritis and degenerative diseases of eye. The antioxidant vegetables have beneficial effects against several of these diseases. Vegetables are most effective against those cancers that involve epithelial cells such as cancers of lung, esophagus, stomach, colon, pancreas and cervix. Vegetables contain compounds such as sulphoraphane that induces GSH transferase, there by helping detoxifying many types of carcinogens.

In view of the above facts, some antioxidant vegetables (including spices and oils), which are commonly taken in the diet have been elucidated in the present article. These include: bhindi, piyaz, lasun, chukandar, sarson, rai, phoolgobhi, pattagobhi, shaljam, lal mirch, bathua, nibu, dhaniya, khira, haldi, gajar, sem, soybean, tamatar, pudina, karela, mitha neem, kamal, French bean, muli, began, alu, palak, methi and adrak. The alcoholic extracts of haldi and karela have been found most effective against metastatic prostate cancer cell lines. Soybean is the contributing factor in the low incidence of breast, prostate, stomach, colon, rectum and lung cancers. Soybean seed contains isoflavonoids, which inhibit the growth of hormone-dependent and hormone-independent cancer cells in culture. The soy isoflavones are the potent inhibitors of cholesterol synthesis, and the plant sterols (*e.g.*, β-sitosterol) and saponins in soy can block cholesterol absorption from the diet or increase cholesterol excretion from the body. Hence, soybean may also protect from the heart disease. Nibu fruit due to its flavonoid, tangeretin and nobiletin contents can potentially inhibits the tumour cell growth and can activate the detoxifying P-450 enzyme system. Vegetables also contain a variety of isoprenoid compounds that exhibit anticancer activity. These compounds include tocotrienols (related to tocopherols) and terpenoids (*e.g.*, limonene, geranoil, menthol, carvone). Overall, these compounds increase tumour latency and decrease tumour multiplicity. These compounds also elicit a significant reduction in total and LDL cholesterol levels, there by reducing the risk of heart disease. There are many vegetable pigments such as flavonoids, carotenoids and anthocyanins. These pigments protect us from various diseases. Quercetin (a flavonoid) chiefly present in red and yellow piyaz and French bean possesses both anticarcinogenic activity and the ability to inhibit LDL oxidation. Anthocyanins and carotenoids reduce the risk of heart disease and cancers. In addition, a variety of phenolic compounds (*e.g.*, caffeic, ellagic and ferulic acids, sesamol, and vanillin) present in vegetables exhibit antioxidant and anticancer activities and inhibit atherosclerosis. Adrak contains phenolic compound known as gingerol, which has antioxidant activity that is even greater than α-tocopherol (vitamin E).

Agriculture Plants Used Against Livestock Diseases

Govind Pandey* and S. Madhuri[1]

Shri Ram Institute of Technology Pharmacy, Jabalpur, M.P., India

SUMMARY

Ethnomedicinal survey exhibits a number of valuable data on indigenous medicinal plants, which are cultivated for agriculture purpose. Many agriculture plants and their products have been used traditionally for the treatment of various livestock diseases. These plants are used not only for food but also employed to cure different ailments of livestock. The medicinal uses of these plants may help veterinarians, medicos and agriculture/bio scientists or researchers in discovering new drugs against various diseases of human and animals. Some of the agriculture plants possessing useful properties against various diseases of livestock are: *Allium cepa, Allium sativum, Annona squamosa, Brassica campestris and B. nigra, B. oleracea* var. *botrytis, Cajanus cajan, Capsicum annum, Carica papaya, Cicer arietinum, Citrus limon, Curcuma longa, Daucus sativa, Elephantopus scaber, Lens culinaris, Lycopersicon esculentum, Momordica charantia, Oryza sativa, Picrorhiza kurroa, Psidium guajava, Raphanus sativus,*

(Accepted : May, 2008).

[Paper published in: International Journal of Plant Sciences, 2008; 3(2): 671-674].

* Correspondence Author: Mobile: 09893769290; E-mail: drgovindpandey@rediffmail.com

Sesamum orientale, Solanum melongena, Solanum tuberosum, Spinacia oleracea, Trachyspermum ammi, Trigonella foenumgraecum, Triticum aestivum, Zea mays and *Zingiber officinale.*

Keywords: *Agriculture plants, Livestock diseases.*

There are about 400 families of the flowering plants in the world, at least 315 are represented by India (Sharma, 2003). Considerable works have been done on ethnomedicinal plants of India. But very little scientific attention has been paid to plants and other natural products, which are specifically used in the treatment of livestock diseases (Tiwari and Pande, 2006). Various agriculture plants and their products have been used in veterinary practices since times immemorial. The dependence of rural mass on the plant-based medicines for curing animals is mainly because of the limited access to the modern (allopathic) medicine system, cheaper and easy availability, and the simplicity of their applications. The herbal drugs also do not possess toxicity or have very less side effects (Jadeja *et al.*, 2006).

Ethnomedicinal survey exhibits a number of valuable data on indigenous medicinal plants, which are cultivated for agriculture purpose. Thus, the present review paper signifies the efficacy of some agriculture plants used against various diseases of livestock. Cattle, buffalo, sheep and goat are the main livestock and domestic animals. The attempt has also been made to improve the knowledge among rural people and farmers for documentation of ethnoveterinary practices.

Some Agriculture Plants Used against Livestock Diseases

The data on 30 agriculture-medicinal plants have been collected from various papers, journals and books, and enumerated below. In general, the main active components, curative uses and some important biological activities of these plants have been mentioned by many authors (Chopra *et al.*, 2002; C.S.I.R., 1986; Kaushik and Dhiman, 1999; Nadkarni, 1954; Sharma, 2003). However, some of these plants have also been reported by some other investigators to be used against specific veterinary diseases as cited under these plants.

Allium cepa Linn. (Piyaz)

Bulb contains amino acids, vitamins (C, E, folic acid), quercetin flavonoid and allicin. Boiled bulb is given after parturition to clean stomach

(Jadeja *et al.,* 2006). The bulbs are also given in flatulence, indigestion, cough, cold, blisters, boils, swelling, pain and fever, and have analgesic, antipyretic, antiinflammatory, diuretic, antidiabetic, anitcholesterol, anticancer and bactericidal activities.

Allium sativum **Linn. (Lasun)**

Bulb contains allicin, allin and amino acids. The bulbs are given in skin and lung disorders, flatulence, colic and dyspepsia, and have antidiabetic activity.

Annona squamosa **Linn. (Sitaphal)**

Plant contains annoaine, higenamine and amino acids. Leaves and seeds are externally applied in wounds, boils, burns and swelling (Jadeja *et al.,* 2006). The seeds are also used for killing external parasites (lice).

Brassica campestris **Linn. (Sarson)**

Seed oil contains glycerides of palmitic, stearic, oleic and linoleic acids. The oil is given in cough and the seeds powder is given in urinary problems (Jadeja *et al.,* 2006). Oil is also used in scabies, external parasites, rheumatism, cancer and ulcer.

Brassica juncea **(*B. nigra*) Linn. (Rai)**

Plant contains essential oils, and seed contains 2 antithiamines and flavonol glycosides. Oil along with rock salt is used in the diseases of gum (Jadeja *et al.,* 2006). Seed powder is applied on forehead to relief pain, while the seeds and oil are used in rheumatism, paralysis, skin diseases, external parasites and dengue fever.

Brassica oleracea **var. *botrytis* Linn. (Phoolgobhi)**

Leaf contains vitamin C, while phoolgobhi contains cysteine sulfoxide and vitamin U. The leaves are externally used in gout, rheumatism, skin diseases and blister. Phoolgobhi is cardiotonic, antiinflammatory and antitussive, and is used in biliousness and urinary diseases.

Cajanus cajan **Linn. (Arhar)**

Leaf contains cajanol, and root contains cajanone and cajaquinone. Roots and yellow leaves in equal quantities are burnt and ash is applied on affected parts in sinus fistula. The expressed juice of leaves is given with salt in jaundice.

***Capsicum annum* Linn. (Lalmirch)**

Fruit contains amino acids, ascorbic and folic acids, glutathione and flavonoids. Fruit powder with oil cures foot diseases (Jadeja *et al.,* 2006). The fruits are stimulant and stomachic, and are used in dyspepsia, neuralgia, rheumatism, sore throat, cough, yellow fever, diarrhoea, snakebite and internal parasites.

***Carica papaya* Linn. (Papita)**

Fruit contains papin and cryptoflavin, while seed contains carpasemine. The seeds and fruits are used in hepatitis, diarrhoea and dysentery. The fruits are digestive, stomachic, carminative, diuretic and anthelmintic.

***Cicer arietinum* Linn. (Chana)**

Seed contains isoliquiritigenin, while plant contains amino acids and vitamins A, D and E. Fresh leaf juice is given to livestock in constipation (Jadeja *et al.,* 2006; Tiwari and Pande, 2006). The macerated seeds in water are used as tonic. The seeds are astringent, diuretic and antibilious, and are given in debility, scurvy and jaundice.

***Citrus limon* Linn. (Baranibu)**

Leaf contains bergapten, citropten and limettin, and fruit contains vitamin C. The leaves are warmed with ghee and tied on the eyes to cure swelling. The fruits are antiseptic, appetizing, astringent and stomachic, and are given in scurvy, rheumatism, dysentery, diarrhoea and liver disorders.

***Curcuma longa* Linn. (Haldi)**

Rhizome contains curcumin, zingiberine and curcuminoids. The rhizomes ground with lalmirch and chhipi root are given with water to cure cough-cold and to expel internal parasites (Jadeja *et al.,* 2006; Jadhav, 2006; Tiwari and Pande, 2006). The rhizome powder is used in sprain, pain, swelling, wound, mastitis, tumour and liver diseases.

***Daucus sativa* DC. (Gajar)**

Root contains pyruvate kinase, acyltransferase and alkane, while seed contains sitosterol and amino acids. Decoction of leaves and seeds is used as stimulant to uterus during parturition. Boiled roots are given to cattle to fatten them. The roots and seeds are stimulant and carminative, and are used in dropsy and kidney diseases.

***Elephantopus scaber* Linn. (Gobhi)**

Decoction of roots and leaves are given in dysuria, diarrhoea, dysentery, swelling or pain in stomach, ulcer and eczema. The whole plant is astringent and cardiac tonic, and is given in snakebite.

***Lens culinaris* Medic. (Masur)**

Plant contains lenticin, tricetin and luteolin, while seed contains protein. Masur gruel is given in haematuria, and paste of seeds is applied on the wounds and broken horns (Tiwari and Pande, 2006). The seeds are laxative and given in constipation.

***Lycopersicon esculentum* Mill. (Tamatar)**

Fruit contains oxalic, citric and mallic acids, vitamins A and C, while seed and herbage contain solanine. The fruits are used for maturation of abscesses. The pulp and juice of fruits are intestinal antiseptic, blood purifier, cholagogue, stimulant and digestive, and are given in pulmonary diseases, dyspepsia and stomatitis.

***Momordica charantia* Linn. (Karela)**

Leaf and fruit contain linoleic and palmitic acids, momordin, proteolytic enzymes and vitamins A, B and C. The leaf juice is externally applied in burning of feet soles and round the eyes to cure night blindness (Sharma, 2003). The fruits and seeds are anthelmintic, carminative and stomachic, and are used in rheumatism, gout and diseases of liver and spleen.

***Oryza sativa* Linn. (Chaaval)**

Silver skin of grain contains oryzanin, and bran contains a glucoside nukain, which yields aglucone nukagenin. Chaaval with dal is used as a perfect food in dysentery and can alone support life as it contains protein, carbohydrate, fat, vitamins and minerals. Decoction of grains is used internally as antiinflammatory for digestive tract.

***Picrorhiza kurroa* R. ex B. (Kutki)**

Fruit contains kutkin and kurrin. The fruits are given in loss of appetite, dropsy, fever, foot and mouth disease (FMD) and internal parasites (Jadeja *et al.*, 2006). Roots are bitter, stomachic and purgative.

***Psidium guajava* Linn. (Amrud)**

Plant contains leucocyanidin and quercetin, while leaf contains eugenol. Decoction of root bark and leaves is given in chronic diarrhoea. The root bark is astringent and fruits are laxative. The leaves are astringent, and applied in wounds and ulcers.

***Raphanus sativus* Linn. (Muli)**

Root contains glucoside, enzymes and methyl mercaptan. The roots are used in dysentery, indigestion, urinary disorders and gastrointestinal pain (Jadeja *et al.,* 2006). Seeds are expectorant, diuretic, laxative and carminative.

***Sesamum orientale* Linn. (Til)**

Seed contains essential oil, proteins sesamin and sesamolin. The seed oil is rubbed on the eyelids or dropped in eye problems. The seeds are diuretic, lactagogue, and are used in urinary complaints and ulcer.

***Solanum melongena* Linn. (Baingan)**

Fruit contains vitamins A, B and C, while seed contains linoleic and oleic acids. The fruits and leaves are given in asthma, weakness, dysuria, bronchitis, liver disorders and ulcers. The seeds are stimulant, and leaves and fruits are sedative and narcotic.

***Solanum tuberosum* Linn. (Alu)**

Plant contains solanine, solasodine and lutein. A portion of raw tuber (alu) is grated over the sore to cure leg ulcers. Alu is also given in weakness, low lactation and cough.

***Spinacia oleracea* Linn. (Palak)**

Plant contains vitamins, iodine, lecithin, carotene and amino acids. The whole plant and leaves are used in inflammation of lung, liver and bowel, fever, jaundice and urinary calculi. Seeds and fruits are laxative, demulcent and diuretic.

***Trachyspermum ammi* Linn. (Ajwain)**

Fruit yields essential oil containing thymol. The fruits are given in weakness after pregnancy and low lactation (Jadhav, 2006). They are also employed in colic, gastric disorders, diarrhoea, cholera, sore throat, bronchitis and pneumonia, and have antiseptic, carminative and stimulant properties.

***Trigonella foenumgraecum* Linn. (Methi)**

Seed contains choline, trigonelline, amino acids, proteins and vitamins, while leaf contains saponins and quercetin. A poultice of seeds is applied to reduce swelling, burns and inflammatory affections (Jadeja *et al.*, 2006). The leaves and seeds are used in rheumatism, cancer, intestinal inflammation, low lactation, indigestion and bilious disorders. The seeds are carminative and tonic.

***Triticum aestivum* Linn. (Gehun)**

Grain contains oligosaccharides, fat and vitamin E. The flour of grains is applied over inflammation and burns (Jadeja *et al.*, 2006), and also given in food poisoning, low lactation and gastric disorders.

***Zea mays* Linn. (Makka)**

Grain contains oxalic acid, carbohydrate, fat, protein and vitamins. Decoction of styles of the plant is given in urinary problems as diuretic. The grains are astringent, nutritive and nourishing.

***Zingiber officinale* Rosc. (Adrak)**

Rhizome yields a volatile oil containing camphene, gingerol, zingiberene, borneol, cineol and proteins. The rhizomes are given with salt before feed to increase appetite and in weakness after pregnancy (Jadeja *et al.*, 2006; Jadhav, 2006). They are also given in swelling, indigestion, vomiting, flatulent colic, diabetes and heart disorders, and are stimulant and carminative.

References

Chopra R.N., S.L. Nayar and I.C. Chopra (2002). Glossary of Indian medicinal plants. Publications and Information Directorate, C.S.I.R., New Delhi.

C.S.I.R. (1986). The useful plants of India. Publications and Information Directorate, C.S.I.R., New Delhi.

Jadeja, B.A., Odedra, N.K., Solanki, K.M. and Baraiya, N.M. (2006). Indigenous animal healthcare practices in district Porbandar, Gujarat. *Indian J. Traditional Knowledge,* 5(2), 253-258.

Jadhav, D. (2006). Ethnomedicinal plants used by Bhil tribe of Bibdod, Madhya Pradesh. *Indian J. Traditional Knowledge,* 5(2): 263-267.

Kaushik, P. and Dhiman, A.K. (1999). Medicinal plants and raw drugs of India. Bishen Singh Mahendra Pal Singh, Dehradun.

Nadkarni, K.M. (1954). Indian Materia Medica (Vol. 2), 3rd Ed. Popular Book Depot, Bombay and Dhootpapeshwar Prakashan Ltd., Bombay.

Sharma, R. (2003). Medicinal Plants of India: An Encyclopedia. Daya Publishing Home, Delhi.

Tiwari, L. and Pande, P.C. (2006). Indigenous veterinary practices of Drama valley of Pithoragarh district, Uttarakhand. *Indian J. Traditional Knowledge,* **5**(2): 201-206.

Some Dietary Agricultural Plants with Anticaner Properties

*S. Madhuri and Govind Pandey**

Faculty of Pharmacy, Shri Ram Institute of Technology, Jabalpur - 482002 (M.P.), India

ABSTRACT

The curative effect of primary stage of cancer is enhanced by combining the dietary supplementation (by agricultural foods) of metabolic deficiency with a simultaneous specific stimulation of the immunity of patients. This combined biological, agricultural and immunological therapy has cured hundreds of cancer patients and experimental animals in randomized studies over 30 years. These plants may promote host resistance against infection by re-stabilizing body equilibrium and conditioning the body tissues. The present article aims to put the knowledge about some dietary agricultural plants and their products, which possess versatile anticancer and immunomodulatory properties. They are commonly consumed in our diet as food matters, and are also helpful for the livelihood of animals. Some dietary agricultural plants which have been reported to possess anticancer properties against various types of malignant (cancer) and benign tumours are: *Allium cepa*

[Paper published in: Plant Archives, 2008; 8(1): 13-16; Indexed in Indian Science Abstracts, 44(14), 2008-013424, and Medicinal and Aromatic Plants Abstracts, 30(6): 875, 2008-06-3380]

****Address for correspondence:*** E-mail: drgovindpandey@rediffmail.com// drmadhurig8@yahoo.co.in; Mobile: 09893769290

(Piyaz), *Allium cepa* var. *aggregatum* (Shallot), *Allium sativum* (Lasun), *Brassica campestris* (*Sarson*), *Brassica oleracea* var. *botrytis* (Phoolgobhi), *Brassica oleracea* var. *capitata* (Pattagobhi), *Brassica rapa* (Shaljam), *Citrus limon* (Baranibu), *Curcuma longa* (Haldi), *Emblica officinalis* (Amla), *Glycine javanica* (Soybean), *Lycopersicon esculentum* (Tamatar), *Momordica charantia* (Karela), *Swertia chirata* (Chirayita), *Trigonella foenumgraecum* (Methi), *Triticum aestivum* (Gehun) and *Zingiber officinale* (Adrak). Some other food plants having anticancer activity include *Amorphophallus companulatus* (Suran), *Avena sativa* (Oat), *Cajanus cajan* (Arhar), *Hordeum vulgare* (Jau), *Lens culinaris* (Masur), *Mentha arvensis* (Podina) and *Zea mays* (Makka).

Keywords: *Cancer, dietary agricultural plants, anticancer properties.*

Introduction

Cancer is considered to be an uncontrolled growth, which is caused by a complex, poorly understood, interplay of genetic and environmental factors. Being chronic in nature it is difficultto cure cancer (malignant tumour) as the cells proliferate to their maximum till it shows clinical symptoms (Pandey and Madhuri, 2006a; Somkuwar, 2003). Since the allopathic system of medicine may cause several untoward effects on the normal host tissues, the indigenous system of medicine (medicinal plants) has become popular throughout the world (Pandey and Madhuri, 2006b). In India, 45,000 plant species have been identified, out of which about 15-20 thousand plants are of good medicinal value. However, traditional communities use only about 7000-7500 plants for medicinal purpose. The Siddha system of medicine uses about 600, Ayurveda 700, Unani 700 and modern medicine about 30 medicinal plants for treating variety of diseases in man and animal. However, only few medicinal plants have attracted the interest of scientists to investigate the remedy for cancer (Somkuwar, 2003).

Use of plants for treating various diseases of both animal and man is old practice. The agricultural plants not only maintain the health and vitality of individuals, but also cure various diseases including cancer without causing any toxicity. According to WHO estimates, more than 80 per cent people in developing countries depend on traditional medicine for their primary health needs, and a recent survey shows that more than 60 per cent of cancer patients use vitamins or herbs as therapy (Madhuri *et al.*, 2007). Vegetables are the most important parts of our food, which are consumed daily. Consumption of fruits and vegetables is widely accepted as lowering the risk of several cancers. The antioxidant fruits and vegetables prevent from the cancer and other diseases by protecting cells from damage. The phytochemicals which show antioxidant activities include vitamins (A, C, E, K), carotenoids, terpenoids, flavonoids, polyphenols, saponins, enzymes and minerals (Kathiresan *et al.*, 2006).

Some agricultural plants and their products such as vegetables, fruits and crops play an important role for the prevention of cancer in our life. Many naturally occurring substances present in the diet/food have been identified as potential chemopreventive agents, and consuming large amount of vegetables and fruits can prevent the development of cancer. Consumption of fruits and vegetables is widely accepted as lowering the risk of most common cancers. Many doctors recommend that people wishing to reduce their risk of cancer eat several pieces of fruits and several portions of vegetables every day. An inverse relationship has been suggested between the consumption of fruits and vegetables and the incidence of cancer in multiple organs including lung, larynx, mouth, pharynx, gastrointestinal tract and pancreas (Rao *et al.*, 2004; Vechhia and Tawani, 1998). Many plant-derived products exhibit potent anticancer activity against several rodent and human cancer cell lines. A recently published study of 2400 Greek women noted that consumption of vegetables and fruits were independently associated with significant reductions in the incidence of breast cancer. Compared with meat eaters, most, but not all, studies have found that vegetarians are less likely to be diagnosed with cancer. Vegetarians have also been shown to have stronger immune function, possibly explaining why vegetarians may be partially protected against cancer (Armstrong *et al.*, 1981).

Scientific Reports on some Dietary Agricultural Plants as Anticancer

In view of the above facts, the present article has been reviewed with the aims to put the knowledge about some dietary agricultural plants and their products, which possess versatile anticancer and immunomodulatory properties. They are commonly consumed in our diet as food matters, and are also helpful for the livelihood of animals. The data on 21 such agricultural plants have been collected from various sources (C.S.I.R., 1986;

I.D.M.A., 2002; Jarald and Jarald, 2006; Kaushik and Dhiman, 1999; Madhuri *et al.*, 2007; Prajapati *et al.*, 2003) that are mentioned here. However, the anticancer properties of these agricultural plants have been specifically reported by some other authors as elucidated under each plant.

1. *Allium cepa* Linn. (Piyaz)

Bulb contains diallyl disulphide (a sulphurous product), quercetin flavonoid, allicin, allin, amino acids and vitamins (folic acid, C, E). Diallyl disulphide inhibits cancerous cells in stomach. The quercetin, because of its antioxidant properties, may treat lung cancer and other cancers (Vecchia and Tawani, 1998; Prajapati *et al.*, 2003).

2. *Allium cepa* var. *aggregatum* (Shallot)

Bulb contains allyl propyl disulphide (a sulphurous product), flavonoid, phenolic content, proteins and vitamins (folic acid, C). Allyl propyl disulphide, phenolic content and flavonoid inhibit cancerous cells in liver, stomach and other organs. Shallots help liver to eliminate toxins to inhibit and kill cancer cells (Cherry, 2006).

3. *Allium sativum* Linn. (Lasun)

Bulb contains sulphur compounds (diallyl sulphide, compounds inhibit cell proliferation, modulate cell cycle activity and interfere with hormone action in cancer cells. The allicin inhibits the proliferation of cancerous cells of human mammary gland, endometrium and colon. Lasun is also used in gastric and liver cancers (Arivazhagan *et al.*, 2004; Heber, 2004; Prajapati *et al.*, 2003).

4. *Brassica campestris* Linn. (Sarson)

Seed oil contains glycerides of palmitic, oleic and linoleic acids. It is used in various tumours (Vecchia and Tawani, 1998).

5. *Brassica oleracea* var. *botrytis* Linn. (Phoolgobhi)

Sprout contains cysteine, ascorbigen, sulphoxide, indole-3-carbinol, glucaric acid, sulphoraphane, glucosinolates, isothiocyanates and vitamins A, C. These components possess anticancer activities in lung, stomach, colon, bladder, mammary and rectum cancers (Beecher, 1994; Heber, 2004).

6. *Brassica oleracea* var. *capitata* Linn. (Pattagobhi)

Cabbage contains glucaric acid, sulphoraphane, glucosinolates, isothiocyanates, cysteine, ascorbigen, sulphoxide, indole-3-carbinol, allyl isothionate and vitamins A, B, C. These have anticancer activities against bladder lung, stomach, colon, rectum and mammary cancers (Beecher, 1994; Heber, 2004).

7. *Brassica rapa* Linn. (Shaljam)

Leaf and root contain ascorbigen and vitamins (A, C), and are used in various tumours (Heber, 2004; Vecchia and Tawani, 1998).

8. *Citrus limon* Linn. (Baranibu)

Fruit contains flavonoid, flavone, limonoid, limonene, nobiletin and tangeretin. The flavonoid, tangeretin and nobiletin are potent inhibitors of tumor cell growth and can activate the detoxifying P-450 enzyme system. Limonoids inhibit tumor formation by stimulating the enzyme glutathione S-transferase (GST). The limonene (a terpenoid) also possesses anticancer activity. Nibu fruit is used for inhibition of human breast cancer cell proliferation and delaying of mammary tumorigenesis. It is also used in metastasis and leukemia (Heber, 2004; Hirano *et al.*, 1995).

9. *Curcuma longa* Linn. (Haldi)

Rhizome contains curcumin, zingiberine and curcuminoids. It is used in colon, bladder and prostate cancers, intravesical tumour, fibrosarcoma, hepatocellular carcinoma, oesophagal carcinogenesis, leukaemia, stomach papilloma and solid tumour cell lines (Aruna and Sivaramkrishanan, 1990; I.D.M.A., 2002; Rao *et al.*, 2004; Singhal *et al.*, 2000).

10. *Emblica officinalis* Gaertn. (Amla)

Fruit contains ascorbic acid (vitamin C) and phyllembic acid. It inhibits cellular mutation and may prevent cancer. Its 18 compounds inhibit the growth of gastric and uterine cancer cells. Amla also inhibited the growth of *in vitro* human breast cancer cells (*Emblica officinalis* as anticancer website, 2006; I.D.M.A., 2002; Lambertini *et al.*, 2004; Zhang *et al.*, 2004).

11. *Glycine javanica* Linn. (Soybean)

Soybean seed contains phytates, protease inhibitors, phytosterols, saponins, isoflavonoids and isoflavones. Soybean is the contributing factor in the low incidence of breast, prostate, stomach, colon, rectum and lung cancers. The isoflavonoids inhibit the growth of hormone-dependent and hormone-independent cancer cells in culture. The isoflavones inhibit the growth of human breast and prostate cancers (Heber, 2004).

12. *Lycopersicon esculentum* Mill. (Tamatar)

Fruits contains lycopene. The lycopene, an antioxidant inhibits the proliferation of cancer cells. Higher intake of tamatar or higher blood levels of lycopene is correlated with protection from cancer; the protective effect was strongest for cancers of prostate, lung and stomach but some protective effects also appeared for cancers of pancreas, colon, rectum, esophagus (throat), mouth, breast and cervix (Giovannucci, 1999; Heber, 2004; Kaushik and Dhiman, 1999; Vecchia and Tawani, 1998).

13. *Momordica charantia* Linn. (Karela)

Leaf, fruit and seed contain linoleic acid, momordin, palmitic acid, proteolytic enzymes and vitamins (A, C). These are used in colon, mammary and bladder carcinomas, lymphoma, leukemia and other tumours (Prajapati *et al.*, 2003; Rao *et al.*, 2004).

14. *Swertia chirata* Buch.-Ham. (Chirayita)

Whole plant contains swertianin, chiratanin, swertinin and swertenol. It is used in various tumours (Mandal *et al.*, 1998).

15. *Trigonella foenumgraecum* Linn. (Methi)

Leaf and seed contain choline, trigonelline, amino acids, proteins, vitamins and quercetin. These are used in various cancers (Aruna and Sivaramkrishnan, 1990; Rao *et al.*, 2004).

16. *Triticum aestivum* Linn. (Gehun)

Grain contains ellagic, linolenic and oleic acids, sterols, phytases, tocotrienols and vitamin E. Whole grains protect against lymphomas and cancers of pancreas, stomach, colon, rectum, breast, uterus, mouth, throat, liver and thyroid (Vecchia and Tawani, 1998).

17. *Zingiber officinale* Linn. (Sonth)

Rhizome contains camphene, gingerol, zingiberene, borneol, cineol and proteins. It is used in various cancers (Aruna and Sivaramkrishnan, 1990; C.S.I.R., 1986; I.D.M.A., 2002; Kaushik and Dhiman, 1999).

Besides above, some other dietary agricultural plants/foods exhibit anticancer properties as reported by several authors (C.S.I.R., 1986; Kaushik and Dhiman, 1999 and Prajapati *et al.*, 2003). They are *Amorphophallus companulatus* (Suran), *Avena sativa* (Oat), *Cajanus cajan* (Arhar), *Hordeum vulgare* (Jau), *Lens culinaris* (Masur), *Mentha arvensis* (Podina) and *Zea mays* (Makka).

Conclusion

Many plants and plant products have been used traditionally for the treatment of cancer. However, only some dietary agricultural plants have been enumerated here, which are most commonly cultivated and used not only for food but are also employed for the prevention and treatment of many cancers. These plants act by stimulating both non-specific and specific immunity Fruits and vegetables are the most important parts of our food, which are consumed daily. Consumption of fruits and vegetables is widely accepted as lowering the risk of several cancers. The antioxidant vegetables prevent from the cancer by protecting cells from damage. Thus, the dietary agricultural plants and their products possess versatile antioxidant and immunomodulatory activities leading to anticancer properties. The pharmaceutical formulations of these plants may help medicos, veterinarians, agriculture researchers, biologists, pharmacologists and pharmaceutical scientists for discovering new drugs against various types of cancers occurring in human and animals.

References

Arivazhagan, S., B. Velmurugan, V. Bhuvaneswari and S. Nagini (2004). Effect of aqueous extract of garlic (*Allium sativum*) and neem (*Azadirachta indica*) leaf containing blood oxidant-antioxidant status during experimental gastric carcinogesis. *J. Med. Food.* 7(3) : 334-339.

Armstrong, B. K., J. B. Brown and H.T. Clarke (1981). Diet and reproductive hormones: a study of vegetarian and nonvegetarian postmenopausal women. *J. Natl. Cancer Inst.* 67 : 761-767.

Aruna, K. and V.M. Sivaramkrishnan (1990). Plant products as protective agents against cancer. *Indian J. Exp. Biol.* 28 : 1008-1011.

Beecher, C.W.W. (1994). Cancer preventive properties of varieties of *Brassica oleracea*: A review. *Am. J. Clin. Nutr.* 59(Suppl) : 1166-1170.

Cherry, A. (2006). Alternative medicine in food-shallots. Meta Suchmaschine- seek XL de- website.

C.S.I.R. (1986). The Useful Plants of India. Publications and Information Directorate, New Delhi.

Emblica officinalis as anticancer (2006)- website.

Giovannucci, E. (1999). Tomatoes, tomato-based products, lycopene, and cancer: Review of the epidemiologic literature. *J. Natl. Cancer Inst.* 91: 317-331.

Heber, D. (2004). Vegetables, fruits and phytoestrogens in the prevention of diseases. *J. Postgrad. Med.* 50 : 145-149.

Hirano, T., K. Abe, M. Gotoh and K. Oka (1995). Citrus flavone tangeretin inhibits leukaemic HL-60 cell growth partially through induction of apoptosis with less cytotoxicity on normal lymphocytes. *Br. J. Cancer.* 82 : 1380-1388.

I.D.M.A. (2002). Indian Herbal Pharmacopoeia. Mumbai.

Jarald, E.E. and Jarald, S.E. (2006). Colour Atlas of Medicinal Plants. New Delhi.

Kathiresan, K., N.S. Boopathy and S. Kavitha (2006). Costal vegetation-An underexplored source of anticancer drugs. *Nat. Prod. Rad.* 5(2) : 115-119.

Kaushik, P. and A.K. Dhiman (1999). Medicinal plants and raw drugs of India. Bishen Singh Mahendra Pal Singh, Dehradun.

Lambertini, E., R. Piva, M.T. Khan, I. Lampronti, N. Bianchi, M. Borgatti and R. Gambari (2004). Effect of Bangladeshi medicinal plants on *in vitro* proliferation of human breast cancer cell lines and expression of estrogen receptor alpha gene. *Int. J. Oncol.* 24(2) : 419-423.

Madhuri, S., Pandey, Govind P. and Gupta, V.B. (2007). Common agriculture plants as anticancer (Abs.). *Abstracts (I):* 314-315 (Paper presented in the International Conference held on 14th-16th February, 2007 at Jabalpur, M.P.).

Mandal, S., R. Jain and S. Mukhopadhyay (1998). Naturally occurring iridoids with pharmacological activity. *Indian J. Pharmace. Sci.* 60 : 123-127.

Pandey, Govind and Madhuri S. (2006 a) Autochthonous herbal products in the treatment of cancer. *Phytomedica.* 7 : 99-104.

Pandey, Govind and Madhuri S. (2006 b). Medicinal plants: Better remedy for neoplasm. *Indian Drug.* 43(11) : 869-874.

Prajapati N.D., S.S. Purohit, A.K. Sharma and T. Kumar (2003). *A Hand Book of Medicinal Plants,* 1st Ed. Agrobios (India).

Rao, K.V.K., A. Stanley, H.K. Nair, R. Aalinkeel, S. Mahajan, R. Chawda and M.P.N. Nair (2004). Plant derived products as a source of cellular growth inhibitory pytochemicals on PC-3M, DU- 145 and LNCaP prostate cacer cell line. *Current Science.* 87 : 1585-1588.

Singhal, S.S., S. Awasthi and U. Pandya (1999). The Effect of curcumin on glutathione linked enzymes in K562 human leuklemia cells. *Toxicol. Lett.* 109 : 87-95.

Sivalokanathan S., M. Ilayaraja and M.P. Balasubramanium (2005). Efficacy of *Terminalia arjuna* (Roxb.) on N-nitrosodiethylamine induced hepatocellular carcinoma in rats. *Ind J Exp Biol.* 43 : 264-267.

Somkuwar A.P. (2003). Studies on anticancer effects of *Ocimum sanctum* and *Withania somnifera* on experimentally induced cancer in mice. Ph.D. thesis,J.N.K.V.V., Jabalpur, M.P.

Vecchia, C.L. and A. Tawani (1998). Fruits, vegetables and human cancer. *Eur. J. Cancer.* 7 : 3-8.

Zhang, Y.J., T. Nagao, T. Tanaka, C.R. Yang, H. Okabe and I. Kouno (2004). Antiproliferative activity of the main constituents from *Phyllanthus emblica. Biol. Pharm. Bull.* 27(2) : 251-255.

Cancer Pathogenesis Due to Xenoestrogens of Environments and Food Contaminations

Govind Pandey

Rinder Pest (Govt. Animal Husbandry/Veterinary Department), Jabalpur Division, Jabalpur, M.P., India

E-mail: drgovindpandey@rediffmail.com; drgovindpandey@yahoo.co.in

Several reports indicate that environments and food contaminations including therapeutic agents, *viz.* antibiotics, antihypertensive drugs and oestrogenic endocrine disruptors (as xenoestrogens), may lead to the growth of many cancers. Oestrogens and their metabolites are involved in the pathogenesis of cancers of breast, uterus, ovary, pituitary gland, testicle, liver, kidney and bone marrow. Xenoestrogens or xenobiotic chemicals that can cause cancer include polychlorinated biphenyl congeners, pesticides (*e.g.*, dieldrin, endosulfan and pentachlorophenol), food-related mycotoxin zearalenone and its derivatives, ultraviolet screen, 3-(4-methylbenzyliden)-camphor, some metals (*e.g.* cadmium), fungicides (*e.g.* fenarimol, procymidone and vinclozolin), algicides (*e.g.* triphenyltin), oestrogens, retinoids, pyrethroid insecticides (*e.g.* sumithrin), pentachlorophenol, β-hexachlorocyclohexane and others (*e.g.* atrazine, DDT, hexachlorophene and toxaphene). Diet has influence on cancer development, and several compounds, either present as dietary

---

(Abstract paper published in: Abstracts of International Symposium and XXIX Annual Conference of Society of Toxicology, India organized by National Institute of Nutrition, Hyderabad, AP held on November 5-7, 2009; p. 47).

components or contaminants or formed during food processing, can play a role in cancer risk. All these can adversely interfere with the physiological functions of oestrogens; mediate the hormonal responses by binding to oestrogen receptors; interact with steroid hormone binding proteins; inhibit oestrogen synthesis; and overall can interact with enzyme systems to modulate the oestrogen metabolism.

Radiations in the Pathogenesis of Cancer

Govind Pandey[1] *and S. Madhuri*[2]

[1]*Ex-Professor of Pharmacology; Presently, Officer-In-Charge of Rinder Pest (Veterinary/Animal Husbandry Department, Govt. of M.P.), Jabalpur Division, Jabalpur, M.P., India*

[2]*Senior Research Fellow of CSIR and Ph.D. Scholar, Department of Pharmacology and Toxicology, College of Veterinary Science and A.H., Jabalpur*

ABSTRACT

Radiation is the emission of energy from any source. High frequency radiation (ionizing radiation and UV radiation) may cause genetic damage, leading to cancer. The ionizing radiations include X-rays, γ-rays, cosmic rays and particles given off by radioactive materials such as α and β particles, protons and neutrons. Radiant energy (UV rays of sunlight, ionizing electromagnetic or particulate radiation) induced the neoplasm (cancer) in both humans and experimental animals. The ionized molecules are unstable and quickly undergo chemical changes, thereby forming the free radicals that can damage the DNA molecule or other molecules around it. Thus, the ionizing radiations can cause mutation (change) in a cell's DNA, resulting into the development of cancer. Thyroid gland and bone marrow are most sensitive to radiations; while kidney, bladder and ovary are least affected. Some forms of leukemia (a cancer of bone marrow) appear to be the most common radiation-induced cancers. Studies on the atomic blasts in Japan demonstrated that high-dose radiation increases the risk of developing several cancers including leukemia. Breast cancer can be developed in

---

(Paper published in: Jigyasa, 2009; 3: 86-90)

elderly women by a much more extensive use of personal computers (more than 3 hours a day), mobile telephones, TV sets and other household electrical appliances. Children appear to be twice as sensitive as adults to the leukemia-causing effects of radiation, and unborn children exposed to radiation in the uterus are even more sensitive. The breast cancer risk is more than twice as high as normal; however, the risk of developing lung cancer is 50 per cent higher and the risk for multiple myeloma is more than twice as high as in the general population. The radiation therapy at high doses can cause DNA mutations in cells that survive the radiation, which can lead to the development of a second primary cancer. Of all the types of non-ionizing radiations, only ultraviolet (UV) rays are cancer-causing agents. The sun is the major source of UV radiation, and the most skin cancers are a direct result of sunlight exposure.

Introduction

Cancer (malignant tumour) is an abnormal growth and proliferation of cells. It is a frightful disease which is uncontrollable and incurable, and may occur at any time at any age in any part of the body. Cancer is caused by a complex, poorly understood interplay of genetic and several environmental factors. It represents the largest cause of mortality in the world and claims over 6 millions (Chabner *et al.*, 2006; Madhuri, 2008; Pandey and Madhuri, 2006). Many studies have implicated several environmental factors including air, water and industrial pollutants, environmental chemicals and radiations etc. in the pathogenesis of various types of cancer (Kumar *et al.*, 2006; Mukherjee *et al.*, 2006; Pandey, 2007).

Radiation is the emission of energy from any source. At present there is hardly any aspect of human welfare in which nuclear radiation does not play an important role. The use of radiation has become an integral part of modern life. Radiation is used for scientific purposes, medical reasons and for power and energy generation. Kumar *et al.* (2006) stated that the high frequency radiations (ionizing radiation and UV radiation) may cause genetic damage, leading to cancer. Radiant energy, whether in the form of the ultraviolet (UV) rays of sunlight or as ionizing electromagnetic and particulate radiations, induced the neoplasm (cancer) in both humans and experimental animals. The ionizing radiations include x-rays, γ-rays, cosmic rays and particles given off by radioactive materials such as α and β particles, protons and neutrons. The ionizing radiations consist of high-energy waves that are able to penetrate cells and can cause ionization. It has been narrated that the ionized molecules are unstable and quickly undergo chemical changes, thereby forming the free radicals that can damage the molecule or other molecules around it. One type of molecule that is sensitive to ionizing radiation is the DNA. Thus, the ionizing radiations can lead to a mutation (change) in a cell's DNA, which could contribute to cancer, or to damage/death of the cell. The amount of damage is related to the dose of radiation received by the cell. Organs differ in their sensitivity to the effects of radiation. The thyroid gland and bone marrow are most sensitive to radiations; while the kidney, bladder, and ovary seem to be least affected. Some forms of leukemia (the cancer of bone marrow) appear to be the most common radiation-induced cancers. Ionizing radiation is an effective way to treat certain kinds of cancer (www.cancer.org). According to Kumar *et al.* (2006), the cancer of thyroid gland follows closely but only in the young persons. In the intermediate category are the cancers of breast, lung and salivary gland.

In view of the above facts, the present paper has been reviewed to discuss various radiations which may cause the pathogenesis of cancer in human beings as well as animals.

Effects of Radiations

Generally, the body health is affected on the basis of the amount and duration of radiation exposure. The health can be reduced or death may occur due to the radiation exposure in some of the following ways (Bhatia, 2008):

1. Stochastic Health Effects

The stochastic effects are associated with long-term, low-level (chronic) exposure to radiation. 'Stochastic' refers to the likelihood that something will happen. Increased levels of exposure make these health effects more likely to occur, but do not influence the type or severity of the effect. Cancer is considered by most people the primary health effect from radiation exposure. The radiations can cause changes in DNA, which are called mutations.

2. Non-stochastic Health Effects

These effects appear due to high levels of radiation exposure and become more severe as the exposure increases. The short-term, high-level exposure is known as 'acute' exposure. Many non-cancerous health effects of radiations are non-stochastic.

3. Other Long-term Health Effects

Other than cancer, the most prominent long-term health effects are mutations. The mutations can be teratogenic or genetic. The teratogenic mutations are caused by exposure of the foetus in uterus, and affect only the individual who was exposed. They can include smaller head or brain size, poorly formed eyes, abnormally slow growth and mental retardation. The genetic mutations are those that can be passed from parent to child (offspring).

Cancer Pathogenesis by Various Radiations

The childhood leukemia was early connected to power-frequent magnetic fields of radiation already in the pioneering work by Wertheimer and Leeper (1979), and more recently Scandinavian scientists have identified an increased risk of malignant brain tumours such as astrocytoma and meningioma (Hardell *et al.*, 2005; Lonn *et al.*, 2004). In addition, a clear association between adult cancers and FM radio broadcasting radiation has been noticed, both in time and location (Hallberg and Johansson, 2005). The breast cancer has been reported (Beniashvili *et al.*, 2005) to be developed in elderly women by a much more extensive use of personal computers (more than 3 hours a day), mobile telephones, TV sets and other household electrical appliances. Among the elderly women who developed breast cancer in the first time frame, 20 per cent were regularly exposed to power-frequent fields. But in the more modern period 51 per cent were so exposed, mainly through the use of personal computers. The authors found a significant influence of electromagnetic fields on the formation of epithelial mammary tumours. Hallberg and Johansson (2005) already pointed out that there is a strong association between the body-resonant non-ionizing radiation (FM radio, 100 MHz) and the existence of malignant melanoma of skin. Since this frequency range has a penetration depth of about 10 cm into the human body there is a suspicion that resonant currents may affect the immune defense system also when it comes to beating cancer cells in lungs. Due to that, it is well motivated to study in detail how the presence and rate of lung cancer have changed in Sweden, U.K. and other countries as this new environmental factor was added.

There is no safe level or threshold of ionizing radiation. Even the exposure to background radiation causes some cancers. Additional exposures cause additional risks. The hazards of exposure to some kinds of radiation were recognized shortly after the discovery of X-ray in 1895. The skin reactions were observed in many people working with early X-ray generators, and by 1902 the first radiation-caused cancer was reported in a skin sore. Within a few years, a large number of such skin cancers had been observed. The first report of leukemia (a type of cancer that arises in the bone marrow) in radiation workers appeared in 1911. Marie Curie, the discoverer of radium, and her daughter are believed to have died of radiation-caused leukemia. Since that time, many studies have confirmed the cancer-causing effects of some types of radiation (www.cancer.org).

The sun is the major source of UV radiation. Cleaver and Crowley (2002) reported that there is ample evidence from epidemiologic studies that UV rays derived from the sun induce an increased incidence of basal cell and squamous cell carcinomas (the most common types of skin cancer), and possibly malignant melanoma of skin. Kumar *et al.* (2006) mentioned that the UV rays have a number of effects on cells, inhibition of cell division, inactivation of enzymes, induction of mutation, and in sufficient dosage, death of cells. Sen (2008) reported that the relationship between UV radiation and cancer of the face, lower lip and other exposed areas (especially in fair-skinned individuals) is considered to be causal for a long period of time, particularly in people with outdoor occupation or rural residents.

The electromagnetic (X-rays and γ-rays) and particulate (α and β particles, protons and neutrons) radiations are all carcinogenic. The evidence is so voluminous that a few examples suffice. Many of the pioneers in the development of X-rays developed skin cancers. Miners of radioactive elements in central Europe and Rocky Mountain region of the United Sates have a tenfold increased incidence of lung cancers. Most telling is the follow-up of survivors of the atomic bombs dropped on Hiroshima and Nagasaki. Initially, there was a marked increase in the incidence of leukemias, principally acute and chronic myelocytic leukemia after an average latent period of about 7 years. Subsequently, the incidence of many solid tumours of breast, colon, thyroid and lung increased with longer latent periods. In humans, there is a hierarchy of vulnerability of different tissues to radiation induced cancers. Most frequent are the leukemias, except chronic lymphocytic leukemia. An increased risk of breast cancer has become apparent decades later among women exposed during childhood to the atomic bomb (Preston *et al.*, 2002).

A marked increase in the incidence of thyroid cancer has also been noted in areas exposed to the fallout from the nuclear power plant accident in Chernobyl in 1986 (Williams, 2003).

Ionizing radiation therapy is an effective way to treat certain kinds of cancer. During radiation therapy, high doses of ionizing radiation are directed at the cancer, resulting in the death of the cancer cells. However, this can lead to DNA mutations in cells that survive the radiation, which can eventually lead to the development of another cancer (called a second primary cancer). An increase in second primary cancers in the area being irradiated has been observed in patients with several types of cancer following radiation therapy and/or chemotherapy. Treatment for Hodgkin disease (a type of lymphoma) often delivers lower radiation doses to many areas of the body. These treated areas include large amounts of normal tissue. Patients with Hodgkin disease who are treated with radiation therapy are at an increased risk for developing second primary tumours. When considering radiation exposure from radiation therapy treatment, the benefits generally outweigh the risks. However, some combinations of radiation therapy and chemotherapy are more risky than others. Children treated with radiation therapy for hereditary retinoblastoma, a malignant eye tumour, are at an increased risk for developing a type of bone cancer called osteosarcoma. Similarly, people who have nevoid basal cell carcinoma syndrome, a type of skin cancer, are at high risk for development of basal cell cancers in irradiated areas (www.cancer.org).

References

1. Bhatia, A.L. (2008). Effect of radiation and gene mediated radiation protection with special reference to melatonin. Proceedings (Part II, Abstracts- Ani. Vet. Fis. Sci.): 1-51 (95th Indian Science Congress of the ISCA, Visakhapatnam, A.P., 3-7 January, 2008).
2. Beniashvili, D., Avinoachm, I., Baasov, D. and Zusman, I. (2005). The role of household electromagnetic fields in the development of

mammary tumors in women : Clinical case-record observations. Med. Sci. Monit., 11: 10-13.

3. Chabner, B.A., Amrein, P.C., Druker, B.J., Michaelson, M.D., Mitsiades, C.S., Goss, P.E., Ryan, D.P., Ramachandra, S., Richardson, P.G., Supko, J.G. and Wilson, W.H. (2006). Antineoplastic agents. In: Goodman and Gilman's The Pharmacological Basis of Therapeutics, 11th edn. (Brunton, L.L., ed.). McGraw-Hill Co., New York. pp. 1315-1403.
4. Cleaver, J.E. and Crowley, E. (2002). UV damage, DNA repair and skin cancinogenesis. Front Biosci., 7: 1024.
5. Hallberg, A. and Johansson, O. (2005). FM broadcasting exposure time and malignant melanoma incidence. Electromag. Biol. Med., 24: 1-8.
6. Hardell, L., Carlberg, M. and Mild, K.H. (2005). Case-control study on cellular and cordless telephones and the risk for acoustic neuroma or meningioma in patients diagnosed 2000-2003. Neuroepidemiology, 25: 120-128.
7. Kumar, V., Abbas, A.K. and Fausto, N. (2006). Neoplasia. In: Pathologic Basis of Disease, 7th edn. Saunders, Elsevier India Pvt. Ltd., New Delhi. pp. 269-342.
8. Lonn, S., Ahlbom, A., Hall, P. and Feychting, M. (2004). Mobile phone use and the risk of acoustic neuroma. Epidemiology, 15: 653-659.
9. Madhuri, S. (2008). Studies on oestrogen induced uterine and ovarian carcinogenesis and effect of ProImmu in Rats. Ph.D. thesis submitted to Rani Durgavati Vishwavidyalaya, Jabalpur.
10. Mukherjee, S., Bidhan, C.K., Ray, S. and Ray, A. (2006). Environmental contaminants in pathogenesis of breast cancer. Indian J. Exp. Biol., 44: 597-617.
11. Pandey, Govind (2007). Role of malnutrition in causing cancer and its safety measures. Diploma in Nutrition and Health Education (DNHE) project report. I.G.N.O.U., New Delhi.
12. Pandey, Govind and Madhuri, S. (2006). Medicinal plants: Better remedy for neoplasm. Indian Drugs, 43(11): 869-874.

13. Preston *et al.* (2002). Radiation effects on breast cancer risk: a pooled analysis of eight cohorts. Radiat. Res., 158: 220.
14. Sen, R. (2008). Oral cancer- The global scenario. Proceedings (Part II, Abstracts- Med. Sci.): 1-26 (95th Indian Science Congress of the ISCA, Visakhapatnam, A.P., 3-7 January, 2008).
15. Wertheimer, N. and Leeper, E. (1979). Electrical wiring configurations and childhood cancer. Am. J. Epidemiol., 109: 273-284.
16. Williams, D. (2003). Chernobyl, 15 years later, correlation of clinical, epidemiological and molecular outcomes. Ann. Endocrinol., 64: 72.
17. www.cancer.org/docroot/PED/content/PED_1_3X_Radiation_Exposer_and_Cancer.asp.www.icucec.org/files/radiationbertell.pdf.

Radiations as Environmental Hazard and Disaster in Cancer Pathogenesis

Govind Pandey*, Madhuri Sharma** and Asha Khanna*

**Ex-Professor of Pharmacology and Officer-In-Charge of Rinder Pest (Animal Husbandry Department, Govt. of M.P.), Jabalpur Division, Jabalpur, M.P.*

***Senior Research Fellow of C.S.I.R. and Ph.D. Scholar, Department of Zoology & Biotechnology, Model Science College, Jabalpur – 482 001, M.P.; E-mail: drmadhurig8@yahoo.co.in*

****Professor and Head, Department of Zoology and Biotechnology, Model Science College, Jabalpur*

Radiation is the emission of energy from any source. High frequency radiation (ionizing radiation and UV radiation) may cause genetic damage leading to cancer. The ionizing radiation includes X-rays, γ-rays, cosmic rays and particles given off by radioactive materials such as α and β particles, protons and neutrons. Radiant energy (UV rays of sunlight, ionizing electromagnetic or particulate radiation) induced the neoplasm (cancer) in both humans and experimental animals. The ionized molecules are unstable and quickly undergo chemical changes, thereby forming the free radicals that can damage the DNA molecule or other molecules around it. Thus, the ionizing radiation can cause mutation (change) in a cell's DNA, resulting into the development of cancer. Thyroid gland and bone marrow are most sensitive to radiation; while kidney, bladder and ovary

(Abstract paper published in: Abstract of National Seminar organized by Hawabagh Women's College, Jabalpur, MP held on January 12-13, 2009; p. 21).

are least affected. Some forms of leukemia (a cancer of bone marrow) appear to be the most common radiation-induced cancers. Studies on the atomic blasts in Japan demonstrated that high-dose radiation increases the risk of developing several cancers including leukemia. Breast cancer can be developed in elderly women by a much more extensive use of personal computers (more than 3 hours a day), mobile telephones, TV sets and other household electrical appliances. Children appear to be twice as sensitive as adults to the leukemia-causing effects of radiation, and unborn children exposed to radiation in the uterus are even more sensitive. The breast cancer risk is more than twice as high as normal; however, the risk of developing lung cancer is 50 per cent higher and the risk for multiple myeloma is more than twice as high as in the general population. The radiation therapy at high doses can cause DNA mutations in cells that survive the radiation, which can lead to the development of a second primary cancer. Of all the types of non-ionizing radiation, only ultraviolet (UV) rays are cancer-causing agents. The sun is the major source of UV radiation, and the most skin cancers are a direct result of sunlight exposure.

Environmental Contaminants Causing Malnutrition and Cancer

Madhuri Sharma* and Govind Pandey**

**Ph.D. Scholar and Senior Research Fellow of C.S.I.R., Department of Zoology & Biotechnology, Model Science College, Jabalpur – 482 001, M.P.*
***Ex-Professor of Pharmacology and Officer-In-Charge of Rinder Pest (Animal Husbandry Deptt., Govt. of M.P.), Jabalpur Division, Jabalpur, M.P.; E-mail: drgovindpandey@rediffmail.com*

Many reports indicate that environmental contaminants or hazards can cause either malnutrition or cancer. In fact, at least 35 per cent of all cancers worldwide are caused by incorrect diet. In the case of colon cancer, malnutrition may account for 80 per cent of the cases. Hence, malnutrition may cause several diseases including cancer in human beings as well as animals. Malnutrition is directly responsible for 300,000 deaths per year in children younger than 5 years in developing countries and contributes indirectly to over half the deaths in childhood worldwide. Cancer is the second leading cause of human deaths in the world and claims over 6 millions. Thus the food eaten must not only be nutritious but also it must be complete and clean without any contamination, otherwise the person eating the food would get ill even the food is nutritious. If the right food is not consumed in right amounts by a person it results in malnutrition. The most important environmental or xenobiotic (environmental or foreign)

(Abstract paper published in: Abstract of National Seminar organized by Hawabagh Women's College, Jabalpur, MP held on January 12-13, 2009; p. 45) chemicals causing malnutrition and cancer include polychlorinated biphenyl congeners, pesticides (*e.g.*, dieldrin, endosulfan and pentachlorophenol), food-related mycotoxin zearalenone and its derivatives, ultraviolet screen, 3-(4-methylbenzyliden)-camphor, some metals (*e.g.*, cadmium), fungicides (*e.g.*, fenarimol, procymidone and vinclozolin), algicides (*e.g.*, triphenyltin), oestrogens (*e.g.*, oestradiol, ethinyl oestradiol and diethylstilbestrol), retinoids, pyrethroid insecticides (*e.g.*, sumithrin), pentachlorophenol, β-hexachlorocyclohexane, and others (*e.g.*, atrazine, DDT, hexachlorophene and toxaphene). Various compounds, either present as dietary components or contaminants or formed during food processing, may increase the risk of cancer.

Malnutrition Leading to Cancer by Some Environmental Hazards

Pandey Govind

Officer-In-Charge of Rinder Pest (Animal Husbandry Department, Govt. of MP), Jabalpur Division, Jabalpur, M.P., India

ABSTRACT

Environmental hazards or contaminants may cause malnutrition or cancer, and malnutrition ultimately causes many diseases, including cancer both in human beings and animals. Malnutrition is directly responsible for 300,000 deaths per year worldwide in children younger than 5 years in developing countries. However, cancer is the second leading cause of human deaths in the world and claims over 6 millions. At least 35 per cent of all cancers worldwide are caused by incorrect diet. In the case of colon cancer, malnutrition may account for 80 per cent of the cancer cases. Hence, the food eaten must not only be nutritious but also it must be complete and clean without any contamination, otherwise the person eating the food would get ill even the food is nutritious. If the right food is not consumed in right amounts by a person, it results in malnutrition. The most important environmental or xenobiotic (foreign) chemicals causing malnutrition and cancer are polychlorinated biphenyl

---

Received: 01-10-2010; Revised: 31-10-2010; Accepted: 11-11-2010.

[Paper published in: International Journal of Research in Ayurveda and Pharmacy, 2010; 1(2): 287-291].

congeners, pesticides, food-related mycotoxin and its derivatives, ultraviolet screen, camphor, some metals, fungicides, algicides, oestrogens, retinoids, pyrethroid insecticides, pentachlorophenol, β-hexachlorocyclohexane, etc.

Keywords: *Cancer, Diet or food, Environmental hazards, Malnutrition, Xenobiotics.*

Introduction

Cancer is an abnormal growth and proliferation of cells. It is a frightful disease because the patient suffers pain, disfigurement and loss of many physiological processes. Cancer may be uncontrollable and incurable, and may occur at any time at any age in any part of the body. It is caused by a complex, poorly understood interplay of genetic and environmental factors. Cancer continues to represent the largest cause of mortality in the world and claims over 6 millions[1–4]. In another report[5], it has been disclosed that the incidence of cancer is increasing worldwide, and it is the second leading cause of death. It has been pointed out[2,6–8] that cancer can be caused in one of the three ways: incorrect diet, genetic predisposition and via the environment. At least 35 per cent of all cancers worldwide are caused by incorrect diet, and in the case of colon cancer, diet may account for 80 per cent of the cases. When one adds alcohol and cigarettes to the diet, the percentage may increase to 60 per cent. The genetic predisposition gives rise to 20 per cent cancer cases. Thus the majority of cancers are being associated with a host of environmental carcinogens.

Nutrition is closely associated with health. If a person eats right kind of foods in right amounts, that person will keep good health provided no other factors intervene. If the right food is not consumed in right amounts by a person it results in 'malnutrition' (mal-noo-trish-un). This malnutrition can result in either inadequate or excessive intake of food[9], which can ultimately lead to many diseases[2,10]. Malnutrition is directly responsible for 300,000 deaths per year in children younger than 5 years in developing countries and contributes indirectly to over half the deaths in childhood worldwide. Globally, with 113.4 million children younger than 5 years affected as measured by low weight for age. The overwhelming majority of these children, 112.8 million, will live in developing countries with 70 per cent of these children in Asia, particularly the south-central region, and 26 per cent in Africa. An additional 165 million (29 per cent) children will have stunted length/height secondary to poor nutrition[2,11].

Several reports indicate that environmental and food hazards or contaminants, including xenobiotics (environmental or foreign chemicals) such as oestrogenic endocrine disruptors (xenoestrogens), certain therapeutic agents (*e.g.*, antibiotics, antihypertensive drugs, etc.) may cause various types of cancer[12]. The 'endocrine disruptors' influence the normal functions of oestrogens, thereby cause the cancers of several organs. Chemicals, industrial wastes, pesticides, oral contraceptives (OCs) or oestrogens, detergents, food additives and plastics, all are sources of environmental toxins and endocrine disruptors. The potential endocrine disrupters such as antibiotics, hormones, plasticizers and non-ionic surfactants are becoming the priority pollutants. Most of these have been shown as carcinogenic[12–13] or they may contaminate the diet or food, thereby causing malnutrition[2].

In view of the above facts, an attempt has been made through this review article to explore out the certain environmental and food contaminants or hazards, leading to the development of many cancers or malnutrition.

Malnutrition and Cancer

Malnutrition refers to both undernutrition and overnutrition. 'Undernutrition' means inadequate intake of right food, while 'overnutrition' means excessive intake of right food[9]. Deficiency of energy and protein may result in 'protein energy malnutrition' (PEM), which can be defined as a range of pathological conditions arising from a deficiency of protein and energy, and is commonly associated with infections[2,14]. This disorder has very serious consequences, leading to cancer and/or death. PEM is the major nutritional problem in India and is widely prevalent among children (0-6 years) but is also observed as starvation in adolescents and adults, mostly lactating women, especially during periods of famine or other emergencies [2,15]. If a person eats right kind of foods in right amounts, that person will keep good health provided no other factors intervene. The food eaten must not only be nutritious but also it must be complete and clean without any contamination, otherwise the person eating the food would get ill even the food is nutritious[2,16].

India is facing a great problem of malnutrition. The malnutrition is the single largest cause to the high rate of infant and child mortality and morbidity in India. The diet surveys showed that about 30 per cent of the families surveyed consumed inadequate amounts of food to provide the necessary nutritional requirements. It was also known that an adequate diet contributed to high mortality and morbidity in the general population particularly in infants and pregnant women. The single major factor responsible for a wide prevalence of malnutrition in India is the poverty. About 50 per cent of our people live below poverty line and even after spending 80 per cent of their income on food, they cannot have a balanced diet[17]. There is an increased risk of malnutrition associated with chronic diseases such as cancer, and diseases of intestinal tract, kidney and liver. Patients with these chronic diseases, especially cancer may lose weight rapidly and become susceptible to undernourishment because they can not absorb valuable vitamins, calories and iron[2].

Malnutrition predisposes a person to infection. On the other hand, infections lead to malnutrition. This inter-relationship and synergistic effect of malnutrition and infections often lead to a high incidence of cancer and/or deaths. The people who take contaminated or malnutritious diet are at greater risk of cancer. In relation to cancer, malnutritional factors act as primary effectors in four situations: carcinogens in food articles, affected bioavailability of nutrients, non-nutritive dietary items and harmful contaminants. The malnutrition acts as carcinogen in many ways to produce cancer in human beings and animals. Nutritional carcinogenesis may occur due to ingestion of toxins, nutritional deficiency or malnutrition, non-bioavailability of the micronutrients and inactivation of the metabolic enzymes (*i.e.*, mixed-function oxidases) present in the liver. Thus the aetiological role of malnutrition precedingclinical cancer has been firmly established[18]. People who are malnourished may be skinny or bloated. Their skin is pale, thick, dry, and bruises easily. Rashes and changes in pigmentation are common. Hair is thin and pulls out easily. Other symptoms of malnutrition include: anaemia, diarrhoea, loss of reflexes and lack of coordination, etc.[19]. All these signs and symptoms are correlated with those of many cancers[2].

Environmental Hazards as Carcinogens

Several reports indicate that environmental contaminants or hazards can cause either malnutrition or cancer. It has been reported[20] that many factors have a major effect on increasing the rates of oral, colon, lung and breast cancers. Some of these factors include increased infections, more use of pesticides, low consumption of fruits and vegetables, increased consumption of alcohol and red meat, more smoking, high industrial pollution, more exposure to sun, decreased physical activity and high occupational exposures. The major portion of chemicals to which humans are exposed is naturally occurring, that are carcinogenic at large doses. Almost every fruit and vegetable contains natural carcinogenic pesticides. Many natural chemicals are ingested as carcinogens from cooking food like roasted coffee. They include caffeic acid, catechol, dichlorodiphenyltrichloroethane (DDT), furfural and benzo(a)pyrene, etc. Food additives (*e.g.*, allyl isothiocyanate, alcohol, butylated hydroxyanisole and saccharin), mycotoxins (*e.g.*, aflatoxin and hepatitis B virus) and synthetic contaminants (*e.g.*, polychlorinated biphenyls and tetyrachlorodibenzo-p-dioxin) are also mutagenic and carcinogenic. Further, aryl hydrocarbons such as dioxins, polychlorinated biphenyls (PCBs) and polycyclic aromatic hydrocarbons (PAHs) have been reported[21] to bind at the cellular aryl hydrocarbon receptor (AhR), leading to the toxicity and carcinogenicity.

Humans are exposed to dioxins (belong to a group of halogenated aromatic hydrocarbons) mainly through contaminated foods. A dioxin compound, 2,3,7,8-tetrachlorodibenzo-para-dioxin (TCDD) is the most toxic congener. The TCDD, benzo(a)pyrene and PCBs can activate AhR, which subsequently induce expression of cytochrome P4501A1 (CYP1A1) and P4501B1 (CYP1B1) enzymes. The oxidative stress caused by induction of cytochrome P450s is one of the toxic effects of dioxin. Pesticides such as insecticides (*viz.*, carbaryl, carbosulfan, DDT, methoxychlor, dieldrin, endosulphan, sumithrin, organophosphorus compounds like malathion and parathion, etc.) and herbicides (*viz.*, alachlor, metolachlor, quarternary ammonium, atrazine, pentachlorophenol, etc.) have become a part of environmental contaminants due to their widespread use in agriculture and disease control. Many pesticides are immunotoxic and found to suppress the cell-mediated immunity[12]. Pesticides such as DDT and alachlor can behave like endogenous oestrogen and function to suppress apoptosis in estrogen receptor (ER)-positive human breast cancer[22]. Wesseling *et al.*[23] reported that some pesticides and related chemicals may act as carcinogens. These xenobiotics adversely affected the lymphocyte function[24], and increased the oxidative stress and lipid peroxidation in various tissues[25].

Epidemiological studies indicate a close association between the process of westernization and an increase in breast cancer incidence. Western countries showed a higher breast cancer risk due to environmental influence. The lifestyles, including oral or dietary habits are responsible for the incidence of many cancers. Diet having several compounds, either as dietary components or contaminants or formed during food processing, has influence on the development of cancer. Mukherjee *et al.*[12] reported that zearalenone (a nonsteroidal compound produced by many *Fusarium* fungi species) can contaminate dairy products and cereals (*e.g.*, barley, corn, maize, rice and wheat). Zearalenone may be a potential promoter of breast tumurigenesis. After consumption, food mutagens undergo metabolic activation or detoxification by different endogenous enzymes. Most mutagens begin their adverse effects at the DNA level by forming DNA adducts with carcinogenic metabolites. The N-nitro compounds, *viz.* N-nitrosamines and nitrosamides (*e.g.*, N-nitrosureas) are a large group of chemicals that are linked with the cancer pathogenesis. Humans are exposed to N-nitrosamines from a variety of foods and tobacco smoke. Some pesticides like atrazine can be converted into genotoxic N-nitrosamines (N-nitrosoatrazine) in the environment or digestive system. PAHs are environmental hazards formed during incomplete combustion of coal, oil, gas, garbage and other organic substances such as tobacco and different food items. Leafy vegetables can be a significant source of PAHs in the human diet. In general, PAHs occur in lower amount in cigarette smoke; human exposure is predominantly from dietary sources. Metabolic activation of PAHs results in DNA binding products. Thus, the westernized dietary habits, including smoking may increase the breast cancer risk. Many lipophilic carcinogens of tobacco smoke, *e.g.*, PAHs like benzo(a)pyrene and N-nitrosamines can be stored in breast tissues. Some PAHs such as 7,12-dimethylbenz(a)athracene (DMBA) and dibenzo(a,l)pyrene are potent mammary carcinogens.

Sen[26] cited that tobacco has unfortunately become a routine part of the personal environment in the world. In India, 40 per cent of total cancer cases are tobacco related. Almost one fourth of the India's population (more than 250 million people) consumes tobacco. More than 10 million children below 15 years are addicted to tobacco. According to WHO, head, neck, lung, throat, urinary bladder, kidney, uterus and colon cancers are mostly caused by tobacco. Bidi is said to be more harmful than cigarette. Smoke tobacco contains more than 4,000 chemical compounds and at least 400 toxins, many of them are known carcinogens and irritants. Some airborne contaminants are PAHs, carbon monoxide and nitrogen dioxide, etc. Tobacco related chewing materials include pan (betal), pan masala, gutka, zarda, snuff, tobacco powder, areca nut and naswar, etc. These substances may increase the risk of oral cancer. Pan masala is unfortunately charged with polyaromatic hydrocarbons, organochlorine pesticides (DDT and BHC), narcotics, metals and minerals. About 15 per cent of oral and pharyngeal cancers may be attributed to dietary deficiencies or malnutrition. Similarly, fungi (*Candida albicans*), viruses (human papilloma virus, hepatitis B virus, human T cell lymphotrophic virus and human immunodeficiency virus), ultraviolet (UV) radiation, immunosuppression, certain occupations (farming and industries) and poor socio-economic status aids in the development of precancerous lesions or carcinoma.

The endocrine disruptors are xenobiotic (environmental/foreign) chemicals that adversely interfere with the natural functions of hormones[12–13]. Oestrogenic endocrine disruptors or xenoestrogens are widely distributed in environment. Several chemicals such as PCB congeners, pesticides (*e.g.*, dieldrin and endosulphan), food-related mycotoxin zearalenone and its derivatives, UV screen and even some metals like cadmium can influence hormonal responses by binding to ER. The xenoestrogens interact with the binding pocket of the ER because they have chemical similarities to oestrogen (usually a phenolic A-ring). Many xenobiotic chemicals can interact with the enzyme systems that metabolize oestrogens; and by this process they may modulate the endogenous metabolism. The AhR or dioxin receptor has involvement with ER-mediated response pathways. Insecticides (*e.g.*, sumithrin) and pentachlorophenol alter steroidsignaling pathways. Other man-made chemicals acting as endocrine disruptors (related with oestrogens) are atrazine, DDT, oestrogens (*e.g.*, oestradiol and ethinyl oestradiol), hexachlorophene and toxaphene, etc. All these can influence the physiological functions of oestrogen and modulate the non-genomic pathways[12], which may cause various cancers.

Acknowledgement

The author is grateful to Dr. Madhuri Sharma, Assistant Professor of Zoology and Biotechnology, NES Science College, Jabalpur, MP for valuable help and support.

References

1. Madhuri S. Studies on oestrogen induced uterine and ovarian carcinogenesis and effect of ProImmu in rat. PhD thesis. Jabalpur, MP, India: RDVV; 2008.
2. Pandey Govind. Role of malnutrition in causing cancer and its safety measures. Diploma in Nutrition and Health Education (DNHE) project report. New Delhi: IGNOU; 2007.
3. Pandey Govind, Madhuri S. Medicinal plants: Better remedy for neoplasm. Indian Drugs 2006; 43(11):869-74.
4. Pandey Govind, Madhuri S. Global scenario of radiation in cancer risk: A review. Int J Pharmaceu Biol Arch 2010; 1(3):195-8.
5. Parkin DM, Bray F, Ferlay J, Pisani P. Global cancer statistics 2002. Cancer J Clin 2005; 55: 74-108.
6. Kumar V, Abbas AK, Fausto N. Neoplasia. In: Pathologic Basis of Disease. 7th ed. New Delhi: Saunders, Elsevier India Pvt Ltd; 2006. p. 269-342.
7. Mohan H. Neoplasia. In: Text Book of Pathology. New Delhi: Jaypee Brothers, Medical Publishers (P) Ltd; 2006. p. 197-240.
8. Reddya L, Odhava B, Bhoolab KD. Natural products for cancer prevention: A global perspective. Pharmacol and Therap 2003; 99:1-13.
9. DNHE-3. Nutrition and health status of the community. In: Nutrition and Health Education. New Delhi: IGNOU; 2004.
10. Copeland EM 3rd, MacFadyen BV Jr, Lanzotti VJ, Dudrick SJ. Intravenous hyperalimentation as an adjunct to cancer chemotherapy. *Am Surg 1975;* 129(2):167-
11. Pupim LB, Kent P, Hakim R. The potential of intradialytic perenteral nutrition: A review. Minor Electrolyte Metab 1999; 25:317-23.

12. Murkherjee S, Bidhan CK, Ray S, Ray A. Environmental contaminants in pathogenesis of breast cancer. Indian J Exp Biol 2006; 44:597-617.
13. Sandhu HS. Pesticides as endocrine disruptors. Souvenir and Abstracts of International Conference of STOX India held at Gwalior, MP on 9-11 October, 2006. pp. 84-85.
14. DNHE-2. Dietary management of disease. In: Public Health and Hygiene. New Delhi: IGNOU; 2003.
15. DNHE-1. Nutrition-related disorders. In: Nutrition for the Community. New Delhi: IGNOU; 2004.
16. Gottschlich MM. The Science and Practice of Nutrition Support. Dubuque: Kendall/Hunt Publishing; 2001.
17. DNHE-3. Nutrition programmes. In: Nutrition and Health Education. New Delhi: IGNOU; 2004.
18. Greenberg DS. Washington report: The Cancer Institute on the Griddle. *N Engl J Med 1978;* 299(4):207-8.
19. Merritt RJ. The ASPEN Nutrition Support Practice Manual. Silver Spring: ASPEN; 1978.
20. Gold LS, Stone TH, Ames BN. Prioritization of possible carcinogenic hazards in food. In: Food Chemical Risk Analysis. New York: Chapman and Hall; 1997. p. 267-95.
21. Kawanishi M, Sakamoto M, Ito A, Kishi K, Yagi T. Construction of reporter yeasts for mouse aryl hydrocarbon receptor ligand activity. Mutat Res 2003; 540:99.
22. Burow ME, Tang Y, Collins-Burow BM, Krajewski S, Reed JC, McLachlan JA, Beckman BS. Effects of environmental estrogens on tumor necrosis factor α-mediated apoptosis in MCF-7 cells. Carcinogenesis 1999; 20:2057.
23. Wesseling C, Antich D, Hogstedt C, Rodriguez AC, Ahlbom A. Geographical difference of cancer incidence in Costa Rica in relation to environmental and occupational pesticide exposure. Int J Epidemiol 1999; 28:365.
24. Banerjee BD, Koner BC, Ray A. Immunotoxicity of pesticides: Perspectives and trends. Indian J Exp Biol 1996; 34:723.

25. Ray S. Studies on paraquat toxicity in rat. PhD thesis. Calcutta, WB, India: University of Calcutta; 2001.

26. Sen R. Oral cancer: The global scenario. Proceedings (Part II, Abstracts-Medical Sciences) of 95th Indian Science Congress of the ISCA held at Visakhapatnam, AP on 3-7 January, 2008. pp. 1-26.

Cancer Pathogenesis Caused by Xenoestrogens of Environment and Food Contaminants: A Review

Govind Pandey[1]* *and S. Madhuri*[2]

[1]*Animal Husbandry Deptt., Jabalpur Division, Jabalpur*

[2]*Department of Zoology and Biotechnology,*

MH College of Home Science and Science for Women, Jabalpur, MP, India

ABSTRACT

The environmental and food contaminants, including therapeutic agents (*e.g.*, antibiotics, antihypertensive drugs, etc.) and oestrogenic endocrine disruptors (xenoestrogens or xenobiotic chemicals), which mediate the oestrogens may lead to the development of cancer. Oestrogens and their metabolites are involved in the cancer pathogenesis of breast, uterus, ovary, pituitary gland, testicle, liver, kidney and bone marrow. Xenoestrogens that can cause cancer include polychlorinated biphenyl congeners, pesticides, food-related mycotoxin zearalenone and its derivatives, ultraviolet screen, some metals, fungicides, algicides, oestrogens, retinoids, pyrethroid insecticides, pentachlorophenol, β-hexachlorocyclohexane, etc. Diet has influence on cancer development, and several compounds, either present as dietary components or contaminants or formed during food processing, may play a role in cancer risk. All these can adversely interfere with the physiological functions of

[Paper published in: Journal of Chemical and Pharmaceutical Research, 2010; 2(4): 687-695].

oestrogens; mediate the hormonal responses by binding to oestrogen receptors; interact with steroid hormone binding proteins; inhibit oestrogen synthesis; and overall can interact with enzyme systems to modulate the oestrogen metabolism.

Keywords: *Environmental and Food contaminants, Xenoestrogens, Cancer.*

Introduction

Cancer is an abnormal growth and proliferation of cells. It is a frightful disease because the patient suffers pain, disfigurement and loss of many physiological processes. Cancer may be uncontrollable and incurable, and may occur at any time at any age in any part of the body. It is caused by a complex, poorly understood interplay of genetic and environmental factors [1-7]. The lack of cell-cell adhesion and increased migration are key characteristics of cancer cells. All cancers occur due to activation or mutation of oncogenes, or inactivation of suppressor genes. More than 20 tumour suppressor genes and 50 oncogenes have been identified and characterized. Most of these genes are involved in activation and detoxification of polycyclic aromatic hydrocarbons (PAHs), suggesting a potential role of these compounds in carcinogenesis [8].

According to the WHO, cancer is the second largest cause of mortality and morbidity in the world after heart disease. 'Cancer' is a word that put an immediate emotional response, which have no relationship to rational thinking, depth of knowledge of the individual and role of individuals in the society. Such responses also occur in doctors, nurses, patients, families, friends of those who suffer from cancer, people who think they may have cancer, teachers, scientists, politicians, social workers and unskilled laboures, etc. Cancer rates continue to rise each year and it kills over 16 million people worldwide per year. However, in India the incidence of cancer may be as high as 3 million cases every year. In addition, more than 12,00,000 deaths occur every year from various types of cancer [8]. As per the other reports [9], cancer is the second leading cause of death in America; colon cancer is the second most common cause of cancer deaths and prostate cancer, second to skin cancer is the most frequently diagnosed cancer among men in the USA; while breast cancer is the most common of cancer in women worldwide. With increase in longevity, it is going to be a problem even in India. Cancers affecting the digestive tract are among the most common of all the cancers associated with aging. The major causes of cancer are smoking, dietary imbalances, hormones and chronic infections leading to chronic inflammation. It has been further stated [10] that cancer is one of the most dreaded diseases, and is leading cause of death; and presently in the USA, cancer accounts for 25 per cent of all deaths in humans. Cancer kills annually about 3500 per million population around the world [2,9,11].

Oestrogens are the most commonly prescribed drugs by far the two major uses are as a component of oral contraceptives (OCs) and hormonal replacement therapy (HRT) in women. OCs used to control the birth, have influenced the lives of untold millions of women since they may cause cancer in humans as well as animals [2,12]. Due to wide use of OCs, a remarkable increase in cases of liver cell adenoma (benign or noncancerous tumour of liver) and hepatocellular carcinoma (HCC) have been reported [13]. There is evidence that long-term use of OCs may increase the risk of breast and cervical cancers. Many studies on animals demonstrate an increased incidence of various types of tumour with oestrogen, hence with the use of OCs. Vaginal, uterine and breast carcinomas have been caused by oestrogens [1-2,14-15]. The association between oestrogen use and cancer has been known for several years. 'Oestrogen cancer' hypothesis postulates that both exogenous and endogenous oestrogens and their metabolites play an etiologic role in the pathogenesis of cancer; and oestrogen administration is followed by regularly reproducible tumours of breast, uterus, ovary, pituitary gland, testicle, kidney and bone marrow either in mouse, rat, rabbit, hamster or dog [16]. Oestrogen can play a tumour-promoting role in target organs. This hormone is responsible for many illnesses in women as well as men [17]. Many reports [18] suggest that oestrogens are closely linked with the pathogenesis of breast cancer. This risk is increased with prolonged exposure by oestrogen; hence, early menarche and/or late menopause increase the risk. The age at first pregnancy, nulliparity, obesity, oestrogen replacement therapy or OCs are also related with the increased risk of breast cancer. The British and American studies [2,14] in women showed that oestrogen increased the risk of uterine (endometrial and cervical) cancer, and the women who begin menstruating early or who start menopause late, produce more oestrogen over their life times and have a higher risk of breast cancer. The natural oestrogens, viz., oestrone and oestradiol are recognized carcinogens in rodents and humans. Oestradiol increased the incidence of tumours of pituitary gland, mammary gland, uterus, cervix, vagina, testes, lymphoid system or bone in various strains of rats and mice [19-20]. In December 2000, the USA Government's National Toxicology Programme added oestrogen to the list of known human carcinogens [21]. It is disconcerting to think that a natural hormone (oestrogen) circulating in significant amounts through the bodies of about half of the world's population (women) is a carcinogen, but it is now official [22]. Ethinyl oestradiol (EO), a semisynthetic 17 β-oestradiol oestrogen, has been reported to cause cytotoxicity, leading to cancer in uterus, ovary and liver after prolonged administration in female albino rats [2,12]. Several reports [23] indicate that environment and food contaminants, including therapeutic agents such as antibiotics and antihypertensive drugs can cause various cancers. Further, the 'endocrine disruptors' influence the normal functions of oestrogens, thereby cause the cancer of several organs. The chemicals, industrial wastes, pesticides, OCs, detergents, food additives and plastics, all are sources of environmental toxins and endocrine disruptors. The potential endocrine disrupters such as antibiotics, hormones, plasticizers and non-ionic surfactants are becoming the priority pollutants; most of these have been shown to be carcinogenic [23-24].

With the above background, an attempt has been made through this review article to explore out certain xenoestrogens of environmental and food contaminants or hazards, leading to the development of many cancers.

Carcinogenic Action of Oestrogen

Oestrogens bind to cytoplasmic estrogen receptors (ERs) in oestrogen responsive tissues. The steroids-receptor complex then translocates to the nucleus, where it brings about changes in transcription. There are two types of ERs, *viz.*, ER α (present in female genital tract, breast, hypothalamus, endothelial cell and vascular smooth muscle) and ER β (present in prostate, ovary and brain). Oestrogens act by interacting with specific ERs in the cytoplasm of target cells; this ER complex then translocates to the nucleus, where it attaches itself to the appropriate gene and mediates the transcription of relevant mRNA [1-2,6]. The differential expression of ER α or β during ovarian carcinogenesis, with overexpression of ER α as compared to ER β in cancer has been demonstrated [25]. This differential expression in ER suggested that oestrogen-induced proteins may act as ovarian tumour-promoting agents. After the oestrogen hormone binds to its receptors in a cell, it turns on hormone-responsive genes that promote DNA synthesis and cell proliferation. If a cell happens to have cancer-causing mutations, those cells will also proliferate and have a chance to grow into tumours. In carcinogenic oestrogen, the reactions produce intermediates capable of producing oxygen radicals that can damage the cell's fats, proteins and DNA. Unrepaired DNA damage can turn into a mutation, leading to cancer [2,12,14].

The oestrogen-induced carcinogenic mechanism has been fully explained [20]. Oestradiol may be converted to 4-hydroxyoestradiol or oestrone. If these catechol oestrogens are not detoxified by phase II enzyme activities such as catechol-O-methyltransferase (COMT), UDP glucuronosyl transferase or sulphotranferase, they may undergo metabolic redox cycling. Quinone or semiquinone intermediates in the redox cycle are free radicals, which generate more oxygen radicals and may induce various types of DNA damage. This genotoxicity may result in gene mutation and cell transformation. Thus, the tumours may develop from cells transformed by genotoxic action of oestrogen, which proliferate in response to an ER-mediated stimulus (Figure 1).

In the metabolism of xenobiotics, cytochrome P450 enzymes or monooxygenases catalyze hydroxylation reaction. This process occurs in two phases: cytochrome P450s are involved in phase I reaction, which sometimes convert biologically inactive compounds into active or toxic metabolites. Subsequently, in phase II, products of phase I reaction are conjugated with many molecules, *e.g.*, glucuronic acid, sulphate, glutathione (GSH), acetyl or methyl groups, leading to excretion from the body. Several phase I and II enzymes are associated with oestrogen metabolism [23]. Oestrogens are mainly hydroxylated by cytochrome P4501A1 (CYP1A1) and CYP1B1 into 2-hydroxyestrogens and 4-hydroxyestrogens, respectively [19,23]. It is believed that 4-hydroxyestrogens may act as a carcinogen [26]. In view of this, Mukherjee *et al.*, [23] has explained the similar carcinogenic mechanism of oestrogen, as proposed by Liehr [20]. The metabolisms of 2- and 4-hydroxyestrogens (catechol oestrogens) are produced in a series of linked oxidation reactions

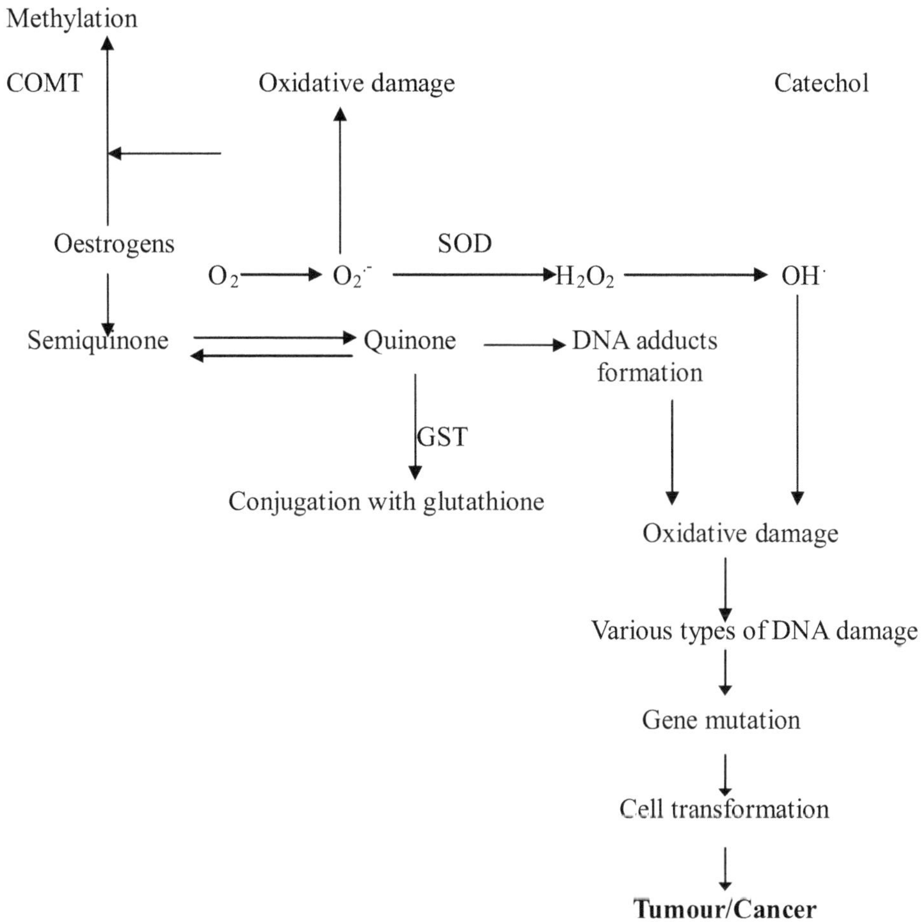

O_2 : Superoxide free radical
OH : Hydroxyl free radical

Figure 1: Proposed carcinogenic action of oestrogen.

that form the oxidative oestrogen metabolism pathway causing cancer. Such metabolism leads to formation of unstable semiquinone, which is an intermediate in both oxidation and reduction reactions, and can react with molecular oxygen to form superoxide radicals and quinone. The superoxide radicals may be reduced to hydrogen peroxide and then to hydroxyl radical in the presence of metal ions. In general, quinones can be conjugated with GSH by glutathione-S-transferase (GST) or can form adducts with guanine and adenine base in DNA. The 2,3-quinone can bind stably to DNA; whereas, 3,4-quinone forms depurinating adducts with guanine and adenine which are lost from DNA by cleavage of the glucosidic bond leaving apurinic sites with mutagenic potential. In addition, quinones and semiquinones undergo redox cycling which results in production of reactive oxygen species (ROS) that can cause oxidative damage to lipids, proteins and DNA. However, several phase II enzymes, *e.g.*, COMT, GST and superoxide dismutase (SOD) participate in catechol metabolism (Figure 1).

Environmental and Food Contaminants as Carcinogens

Several factors have a major effect on increasing the rates of oral, colon, lung and breast cancers, etc. Some of these factors include increased infections, more use of pesticides, low consumption of fruits and vegetables, increased consumption of alcohol and red meat, more smoking, high industrial pollution, more exposure to sun, decreased physical activity and high occupational exposures. The major portion of chemicals to which humans are exposed is naturally occurring, that are carcinogenic at large doses. Almost every fruit and vegetable contains natural carcinogenic pesticides. Many natural chemicals are ingested as carcinogens from cooking food like roasted coffee. These include caffeic acid, chlorogenic acid, catechol, dichlorodiphenyltrichloroethane (DDT), furfural and benzo(a)pyrene, etc. A diet free of naturally occurring carcinogenic chemicals is impossible. Food additives (*e.g.*, allyl isothiocyanate, alcohol, butylated hydroxyanisole and saccharin), mycotoxins (*e.g.*, aflatoxin and hepatitis B virus) and synthetic contaminants (*e.g.*, polychlorinated biphenyls and tetyrachlorodibenzo-p-dioxin) are also mutagenic and carcinogenic [27]. The aryl hydrocarbons such as dioxins, polychlorinated biphenyls (PCBs) and PAHs bind to the cellular aryl hydrocarbon receptor (AhR); and activation of intracellular signaling subsequent to the AhR binding is highly correlated with the toxicity and carcinogenicity of these chemicals [28]. Humans are exposed to dioxins (belong to a group of halogenated aromatic hydrocarbons) mainly through contaminated foods. A dioxin compound, 2,3,7,8-tetrachlorodibenzo-para-dioxin (TCDD) is the most toxic congener. TCDD, benzo(a)pyrene and PCBs can activate AhR, which subsequently induce expression of CYP1A1 and CYP1B1 enzymes. Oestradiol is metabolized by these two enzymes, which activate benzo(a)pyrene to reactive DNA-binding intermediates. There is evidence of cross-talk between ER α and AhR-mediated signaling in breast and endometrial cells, thus the oxidative stress due to induction of cytochrome P450s is one of the toxic effects of dioxin [23].

Pesticides have become a part of environmental contaminants due to their widespread use in agriculture and disease control. Many pesticides are immunotoxic and found to suppress the cell- mediated immunity. Some of the selected pesticides are categorized as under: (a) insecticides such as carbamate (carbaryl, carbosulfan, primicarb, methicarb, etc.), organochlorine (DDT, methoxychlor, aldrin, dieldrin, endosulfan, heptachlor, etc.), organophosphorus compounds (dichlorovos, malathion, parathion, chlopyrifos, trichlorfon, etc.) and pyrethroid (bioallethrin, deltamethrin, sumithrin, etc.); (b) herbicides such as amide (isoxaben, pentachlor, alachlor, metolachlor, etc.), quarternary ammonium (paraquat, diquat, etc.), phenoxy (2,4-dichlorophenoxy acetic acid, 2,4-trichlorophenoxy acetic acid, etc.), trizine (atrazine, simazine, cyanazine, etc.) and unclassified group (oxadiazon, pentachlorophenol, etc.) [23]. The pesticides such as DDT and alachlor can behave like endogenous oestrogen and function to suppress apoptosis in ER-positive human breast cancer [29]. Some pesticides and related chemicals may act as carcinogens [30]. These xenobiotics have adversely affected the lymphocyte function [31] and increase oxidative stress and lipid peroxidation in various tissues [32].

Epidemiological studies indicate a close association between the process of westernization and an increase in breast cancer incidence. Western countries showed a higher breast cancer risk due to environmental influence. Besides reproduction-related factors, a westernized lifestyle is intimately associated with some particular food processing/cooking practices and dietary habits such as frequent use of fast foods. Thus the lifestyles, including oral or dietary habits are responsible for many cancers. Diet having several compounds, either as dietary components or contaminants or formed during food processing, has caused the cancer. Some dairy products such as whole milk and different cheese containing high levels of saturated fat, may increase cancer risk. Milk products also contain growth factor-I, which promotes breast cancer cell growth [33]. The intake of alcohol can increase the cancer risk by enhancing the levels of circulating oestrogen [34]. The food-related mycotoxin zearalenone and its derivatives such as α-zearalanol and β-zearalanol can bind to ER and exert oestrogenic action [35]. Zearalenone, a nonsteroidal compound produced by many *Fusarium* fungi species, can contaminate dairy products and cereals, *e.g.*, barley, corn, maize, rice and wheat. This compound may be a potential promoter of breast tumorigenesis. After consumption, food mutagens undergo metabolic activation or detoxification by different endogenous enzymes. Most mutagens begin their adverse effects at the DNA level by forming DNA adducts with carcinogenic metabolites. Nevertheless, there are some non-genotoxic carcinogens like chloropropanols (usually present in savoury foods). Furthermore, the N-nitro compounds, *viz.*, N-nitrosamines and nitrosamides (*e.g.*, N-nitrosureas) are a large group of chemicals that are linked with the pathogenesis of cancer. Humans are exposed to N-nitrosamines from a variety of foods and tobacco smoke. Interestingly, some pesticides like atrazine can be converted into genotoxic N-nitrosamines (N-nitrosoatrazine) in the environment or digestive system. PAHs are environmental contaminants formed during incomplete combustion of coal, oil, gas, garbage and other organic substances such as tobacco and different food items. Leafy vegetables can be significant sources of PAHs in the human diet. Benzo(a)pyrene is the best PAH compound available from diet. In general, PAHs occur in lower amount in cigarette smoke; human exposure is predominantly from dietary sources. Metabolic activation of PAHs results in DNA binding products. Cooking meat and fish at high temperature produces heterocyclic amines. Generally, heterocyclic amines are formed from creatinine or cretine, amino acids and carbohydrates. Thus, the higher consumption of meat is probably associated with an increased risk of breast cancer [23].

The westernized dietary habits, including smoking may increase the risk of breast cancer. Many lipophilic carcinogens of tobacco smoke, *e.g.*, PAHs like benzo(a)pyrene, 4-aminobiphenyl and N-nitrosamines can be stored in breast tissues. Some PAHs such as 7,12-dimethylbenz(a)athracene (DMBA) and dibenzo(a,l)pyrene are potent mammary carcinogens [23]. Tobacco has unfortunately become a routine part of the personal environment in the world. It is the only legally available consumer product that kills people when it is used as directed. In India, 40 per cent of total cancer cases are tobacco related. Almost one fourth of the India's population consumes tobacco. More than 10 million children below 15 years are addicted to tobacco. WHO links tobacco to 25 cancers, *e.g.*, cancers of head, neck, lung, throat, urinary bladder, kidney, uterus and colon. In crowded environment, *e.g.*, bar and restaurant, second hand smoke (passive smoke) can produce 6 times the pollution of a busy highway. Smoking includes bidi, cigarette, chutta, cheroot, pipe, cigar, dhumti, hookas, kalke, chillum, etc. Bidi has been said to be more harmful than cigarette. Smoke tobacco contains more than 4000 chemical compounds and at least 400 toxins, many of them are known carcinogens and irritants. Some airborne contaminants are PAHs, carbon monoxide, nitrogen dioxide and others. Tar (a mixture of many chemicals including formaldehyde, arsenic, N-nitrosamine, cyanide, benzopyrene, benzene, toluene and acrolein) and heavy metals (cadmium, arsenic, lead and nickel) may cause tobacco cancer. Many cancer promoters, initiators and accelerators as smoking agents are very potent carcinogens. These substances include nuclear aromatic hydrocarbons, chlorostilbenes, catechols, phenols and their substitutes, nitrosonornicotine, nicotine, cotinine nitrosamine and hydrazide [8].

Chewing materials of human beings include pan (betal); pan masala (a combination of areca nut, catechu, tobacco powder, aromatics, lime and brown or grey colouring agents); gutka; pan with tobacco (zarda, dokta or without tobacco); khaini; snuff; tobacco powder; areca nut; and naswar (a local preparation of Afghanistan commercially available as green tobacco powder mixed with calcium carbonate, calcium chloride, ammonium chloride, tobacco ignition residue, sand and other mineral substances). All these substances may increase the risk of oral cancer. Pan masala is unfortunately charged with polyaromatic hydrocarbons (especially dangerous proportion of magnesium carbonate), organochlorine pesticides (DDT and BHC), narcotics, metals and minerals. Heavy and regular consumption of alcohol is a significant risk factor for oral and other cancers. About 15 per cent of oral and pharyngeal cancers may be attributed to dietary deficiencies or imbalances. Similarly, fungi (*Candida albicans*), viruses (human papilloma virus, hepatitis B virus, human T cell lymphotrophic virus and human immunodeficiency virus), UV radiation, immunosuppression, certain occupations (farming and industries) and poor socio-economic status may develop the precancer lesions or carcinoma [8].

Endocrine Disruptors (Xenoestrogens) as Carcinogens

'Endocrine disruptors' are xenobiotic (environmental/foreign) chemicals that adversely interfere with the natural functions of hormones [23-24]. Oestrogenic endocrine disruptors or xenoestrogen are widely distributed in the environment. Several chemicals such as polychlorinated biphenyl congeners, pesticides (*e.g.*, dieldrin, endosulphan, pentachlorophenol), food-related mycotoxin zearalenone and its derivatives, ultraviolet screen, 3-(4-methylbenzyliden)-camphor and even some metals like cadmium can influence hormonal responses by binding to ER. The xenoestrogens interact with the binding pocket of the ER because they have chemical similarities to estrogen (usually a phenolic A-ring). Many xenobiotics can interact with the enzyme systems that metabolize oestrogens; and by this process they may modulate the endogenous metabolism. Numerous chemicals such as fungicides (*e.g.*, fenarimol, procymidone, vinclozolin) and algicides (*e.g.*, triphenyltin) inhibit steroid hormone synthesis. Synthetic compound like diethylstilbestrol (DES) interfere the bio-availability and overall functions of oestrogen by interacting with steroid hormone binding proteins in the blood. The AhR or dioxin receptor has involvement with ER-mediated response pathways. Chemicals like retinoids, pyrethroid insecticides (*e.g.*, sumithrin), pentachlorophenol and β-hexachlorocyclohexane alter steroid signaling pathways. Other man-made chemicals which can act as endocrine disruptors (related with oestrogens) include atrazine, DDT, EO, hexachlorophene and toxaphene, etc. These can influence the oestrogen functions, such as adverse effects on release and excretion of hormones, disruption of regulatory feedback relationships between two endocrine organs, and modulation of non-genomic pathways [23].

References

1. DS Loose; GM Stancel. Estrogens and progestins. In: Goodman and Gilman's The Pharmacological Basis of Therapeutics, 11th Edition (LL Brunton, ed.), McGraw-Hill Co., New York, 2006, 1541-1571.

2. S Madhuri. Studies on oestrogen induced uterine and ovarian carcinogenesis and effect of ProImmu in rat. PhD thesis, RDVV, Jabalpur, MP, 2008.

3. Govind Pandey; S Madhuri. *Phytomedica,* 2006, 7, 98-104.
4. Govind Pandey; S Madhuri. *Biosci. Biotech. Res. Comm.,* 2008, 1(1), 11-18.
5. Govind Pandey; S Madhuri. *Int. J. Pharmaceu. Biol. Arch.,* 2010, 1(3), 195-198.
6. RS Satoskar; SD Bhandarkar; SS Ainapure. Pharmacology and Pharmacotherapeutics, 8th Edition, Popular Prakashan, Mumbai, 2005, 911-934.
7. AP Somkumar. Studies on anticancer effects of *Ocimum sanctum* and *Withania somnifera* on experimentally induced cancer in mice. PhD thesis, JNKVV, Jabalpur, 2003.
8. R Sen. *Proceedings (Part II, Medical Sciences) of 95th Indian Science Congress of the ISCA,* Visakhapatnam, AP, 2008, 1-26.
9. S Madhuri; Govind Pandey. *Current Science,* 2009, 96(6), 779-783.
10. P Balachandran; R Govindarajan. *Pharmacol. Res.,* 2005, 51, 19-30.
11. K Kathiresan; NS Boopathy; S Kavitha. *Nat. Prod. Rad.,* 2006, 5, 115-119.
12. S Madhuri; Govind Pandey. *biomed,* 2010, 5(1), 57-62.
13. HA Edmondson; B Henderson; A Benton. *N. Engl. J. Med.,* 1976, 294, 470-472.
14. Estrogen and cancer website, 2006; www.womenshealth.com, www.amazon.com
15. S Madhuri; Govind Pandey; A Khanna. *Natl. J. Life Sci.,* 2007, 4(1), 67-70.
16. R Hertz. *Cancer,* 1976, 38(1), 534-540.
17. WE Lucas. *Cancer,* 1981, 48(2), 451-454.
18. BK Sharma; A Ray; S Kaur; S Gupta. *Indian J. Exp. Biol.,* 1999, 37, 223.
19. JM Liehr. *Endocrine Review,* 2000, 21(1), 40-54.
20. JM Liehr. *Human Repro. Update,* 2001, 7(3), 273-281.

21. NTP/NIEHS (National Institute of Environmental Health Sciences press release), 2001 (December 15, 2000); www.niehs.nih.gov/oc/news/rocrslt.htm.
22. ME Platt. 2005; We Publish Books at www.book 1234.net.
23. S Murkherjee; CK Bidhan; S Ray; A Ray. *Indian J. Exp. Biol.,* 2006, 44, 597-617.
24. HS Sandhu. *Souvenir and Abstracts of International Conference of STOX, India,* Gwalior, MP, 2006, 84-85.
25. S Cunat; P Hoffmann; P Pujol. *Gynecologic Oncology,* 2004, 94 (1), 25-32.
26. T Rylander-Rudqvist; S Wedren; F Granath; K Humphreys; S Ahlberg; E Weiderpass; M Oscarson; M Ingelman-Sundberg; I Persson. *Carcinogenesis,* 2003, 24, 1533.
27. LS Gold; TH Stone; BN Ames. Prioritization of possible carcinogenic hazards in food. In: *Food Chemical Risk Analysis,* Chapman and Hall, New York, 1997, 267-295.
28. M Kawanishi; M Sakamoto; A Ito; K Kishi; T Yagi. *Mutat. Res.,* 2003, 540, 99.
29. ME Burow; Y Tang; BM Collins-Burow; S Krajewski; JC Reed; JA McLachlan; BS Beckman. *Carcinogenesis,* 1999, 20, 2057.
30. C Wesseling; D Antich; C Hogstedt; AC Rodriguez; A Ahlbom. *Int. J. Epidemiol.,* 1999, 28, 365.
31. BD Banerjee; BC Koner; A Ray. *Indian J. Exp. Biol.,* 1996, 34, 723.
32. S Ray. Studies on paraquat toxicity in rat. PhD thesis, University of Calcutta, 2001.
33. PG Moorman; PD Terry. *Am. J. Clin. Nutr.,* 2004, 80, 5.
34. NC Onland-Moret; PH Peeters; YT van der Schouw; DE Grobbee; CH van Gills. *J. Clin. Endocrinol. Metab.,* 2005, 90, 1414.
35. F Minervini; A Giannoccaro; A Cavallini; A Visconti. *Toxicol. Lett.,* 2005, 159, 272.

Significance of Fruits and Vegetables in Malnutrition Cancer

Govind Pandey[1*] *and S. Madhuri*[2]

[1]*Ex-Professor of Pharmacology (Pharmacy); presently, Officer-in-Charge of Rinder Pest (Department of Veterinary/Animal Husbandry, Govt. of MP), Jabalpur Division, Jabalpur (M.P.), India*

[2]*Assistant Professor of Zoology and Biotechnology, NES College of Science and Commerce, Jabalpur, (M.P.), India*

ABSTRACT

Cancer is the second leading cause of human deaths in the world and claims over 6 millions. In fact, at least 35 per cent of all cancers worldwide are caused by incorrect diet. Malnutrition may cause several diseases, including cancer. Plants have original place in the treatment of cancer with no ill effect. It is distinct that medicinal plants, including fruits and vegetables have versatile immunomodulatory and anticancer activities. Consumption of large amounts of fruits and vegetables can prevent the development of cancer. Many doctors recommend that people wishing to reduce risk of cancer eat several pieces of fruits and several portions of vegetables every day. With this back ground, an attempt has been made to review on certain fruits and vegetables useful against various types of cancer.

Keywords: *Fruits and vegetables, Malnutrition cancer and its treatment, Medicinal plants.*

[Paper published in: Plant Archives, 2010; 10(2): 517-522].

Introduction

Cancer continues to represent the largest cause of mortality in the world and claims over 6 millions, and it is the second leading cause of death (Madhuri, 2008; Pandey, 2007; Pandey and Madhuri, 2010). In fact, cancer may be caused in one of the three ways: incorrect diet, genetic predisposition and via the environment. At least 35 per cent of all cancers worldwide are caused by incorrect diet, and in the case of colon cancer, diet may account for 80 per cent of the cases. When one adds alcohol and cigarettes to diet, the percentage may increase to 60 per cent. The genetic predisposition gives rise to 20 per cent cancer cases. Thus the majority of cancers are associated with a host of environmental carcinogens (Kumar *et al.,* 2006; Madhuri and Pandey, 2009a; Pandey, 2007; Pandey and Madhuri, 2008a; Reddya *et al.,* 2003).

Nutrition is closely associated with health. If a person eats right kind of foods in right amounts, that person will keep good health provided no other factors intervene. If the right food is not consumed in right amounts by a person it results in "malnutrition" (mal-noo-trish-un). This malnutrition can result in either inadequate or excessive intake of food (DNHE-3, 2004a; Madhuri and Pandey, 2009a). An individual will experience malnutrition if the appropriate amount of, or quality of nutrients are not consumed for an extended period of time, which can result in starvation or many other diseases such as scurvy and cancer (Copeland *et al.,* 1975). Malnutrition is directly responsible for 300,000 deaths per year in children younger than 5 years in developing countries and contributes indirectly to over half the deaths in childhood worldwide. Globally, with 113.4 million children younger than 5 years affected as measured by low weight for age. The overwhelming majority of these children, 112.8 million, will live in developing countries with 70 per cent of these children in Asia, particularly the south-central region, and 26 per cent in Africa. An additional 165 million (29 per cent) children will have stunted length/height secondary to poor nutrition (Madhuri and Pandey, 2009 a; Pandey, 2007; Pupim *et al.,* 1999).

Treatment of malnutrition may include increasing the calories and nutrients in diet. Vitamin or mineral supplements can be given if the body is low in vitamins and minerals. Although eating disorders (malnutrition) are treatable and many people recovered, but the recovery is a complex process that can take several months or even years (Pandey, 2007; Parkin *et al.*, 2005). Modern (allopathic) drugs being a major treatment modality for the treatment of diseases, exhibit severe toxicity on normal tissues. Thus there is a definite need to search alternate drugs to be effective in the prevention and cure of diseases caused by malnutrition. Because of variations in the aetiology of cytotoxicity and cancer, there is a great difficulty to diagnose them, and a physician is rarely able to provide any specific treatment. At the most, supportive and symptomatic treatments are given but the multiplicity of deranged functions renders the treatment still more complicated (Madhuri, 2008; Pandey, 1990; Pandey and Madhuri, 2006a).

The plant kingdom plays a major role in the life of human beings and animals. It serves as a food and medicinal sources, and thus maintains the health and vitality of the individuals. Plants have original place in the treatment of diseases with no ill effect. India is the largest producer of medicinal plants and is rightly called the "Botanical garden of the World". In the indigenous or traditional system of medicine, numerous medicinal plants and their formulations are now being used for the treatment and control of various diseases, including cancer (Madhuri, 2008; Madhuri and Pandey, 2009 b; Pandey and Madhuri, 2008a; Parkin *et al.*, 2005; Somkuwar, 2003). More than 50 per cent of all modern drugs in clinical use are of natural products, many of which have the ability to include apoptosis in various tumour cells of human origin (Rosangkima and Prasad, 2004). According to WHO estimates, more than 80 per cent people in developing countries depend on traditional medicine for their primary health needs. A recent survey shows that more than 60 per cent of patients use vitamins or herbs in their therapy (Madhuri and Pandey, 2008; Sivalokanathan *et al.*, 2005). Medicinal plants are known to have good immunomodulatory properties. These act by stimulating both non-specific and specific immunity. They may promote host resistance against infection by re-stabilizing body equilibrium and conditioning the body tissues. Thus, it is distinct that medicinal plants, including fruits and vegetables exhibit immunomodulatory and anticancer activities (Agrawala *et al.*, 2001; Madhuri, 2008; Pandey and Madhuri, 2006b and 2008b).

In view of the above facts, an attempt has been made to explore out the significance of fruits and vegetables for the prevention and treatment of malnutrition cancer.

Malnutrition and Cancer

Cancer is a malignant tumour, which is a disease of human and other multicellular animals. It may be uncontrollable and incurable, and may occur at any time at any age in any part of the body. On the other hand, cancer is a frightful disease because the patient suffers pain, disfigurement and loss of many physiological processes. It is caused by a complex, poorly understood interplay of genetic and environmental factors. Being chronic in nature it is difficult to cure cancer as the cells proliferate to their maximum (Madhuri, 2008; Somkuwar, 2003).

Malnutrition refers to both undernutrition and overnutrition. 'Undernutrition' means inadequate intake of right food, while 'overnutrition' means excessive intake of right food (DNHE-3, 2004a). Due to deficiency of energy and protein, "protein energy malnutrition" (PEM) results, which can be defined as a range of pathological conditions arising from the deficiency of protein and energy, and is commonly associated with infections (Madhuri, 2008; DNHE-2, 2003). This disorder has very serious consequences leading to cancer and/or death. PEM is the major nutritional problem in India and is widely prevalent among children (0-6 years) but is also observed as starvation in adolescents and adults, mostly lactating women, especially during periods of famine or other emergencies (DNHE-1, 2004). If a person eats right kind of foods in right amounts, that person will keep good health provided no other factors intervene. If there is any imbalance in eating pattern (*i.e.*, eating too little or too much), it will result in poor health. The food eaten must not only be nutritious but also it must be complete and clean, otherwise the person eating the food would get ill even the food is nutritious. Human body requires energy to carry out the different activities, and proteins are necessary for growth and repair. The deficiency of these two basic nutrients in two basic nutrients in our body leads to PEM (Madhuri, 2008; Madhuri and Pandey, 2009a; Gottschlich, 2001).

India is facing a great problem of malnutrition. The malnutrition is the single largest cause to the high rate of infant and child mortality and morbidity in India. This condition is not a new health problem in India. As far back as 1946, when the Bhore Committee reviewed the health status of the Indian population, diet surveys had already been done. These surveys showed that about 30 per cent of the families surveyed consumed inadequate amounts of food to provide the necessary nutritional requirements. At that time, it was also known that an adequate diet contributed to high mortality and morbidity in the general population, particularly in infants and pregnant women. The single major factor responsible for malnutrition in India is the poverty. About 50 per cent of our people live below poverty line and even after spending 80 per cent of their income on food, they cannot have a balanced diet. To combat malnutrition, various supplementary nutrition programmes have evolved as short-term measures (DNHE-3, 2004b; Pandey, 2007).

Malnutrition predisposes a person to infection. On the other hand, infections lead to malnutrition. This inter-relationship and synergistic effect of malnutrition and infections often lead to a high incidence of cancer and/or deaths. The people who are at greater risk of having malnutrition include: infants and children who do not get the right type or amount of calories and nutrients; older adults; people who are taking medicines that decrease appetite or affect the digestion and absorption of nutrients; people who abuse drugs or alcohol; people with eating disorders; people with certain diseases such as cystic fibrosis, celiac disease, liver disease, kidney disease and cancer; people with a low income who have trouble to buy the right kind of food; and pregnant women (DNHE-1, 2004; Pandey, 2007). Hence, malnutrition may cause several diseases, including cancer. In relation to cancer, malnutrition needs to be defined comprehensively as the carcinogenic effects produced by nutritional variables through multiple endogenous involvements. Nutritional factors act as primary effectors in four situations: carcinogens in food articles, affected bioavailability of nutrients, non-nutritive dietary items and harmful contaminants. Malnutrition acts as carcinogen in many ways to produce cancer in human beings and animals. Nutritional carcinogenesis may occur due to ingestion of toxins, nutritional deficiency or malnutrition, non-bioavailability of micronutrients and inactivation of metabolic enzymes (mixed-function oxidases) present in the liver. Thus the aetiological role of malnutrition preceding clinical cancer has been firmly established (Greenberg, 1978).

It has been further stated (DNHE-2, 2003) that the abnormal diets such as fad or limited diet and recent decrease in food intake usually cause malnutrition, which may lead to the development of cancer. Poverty and lack of food are the primary reasons why malnutrition occurs in the United States. Ten per cent of all members of low income households do not always have enough healthful food to eat, and malnutrition affects one in four elderly Americans. There is an increased risk of malnutrition associated with chronic diseases such as cancer, and diseases of intestinal tract, kidney and liver. Patients with these chronic diseases, especially cancer may lose weight rapidly and become susceptible to undernourishment because they can not absorb valuable vitamins, calories and iron (Pandey, 2007).

People with drug or alcohol dependencies are also at increased risk of malnutrition. These people tend to maintain inadequate diets for long periods and their ability to absorb nutrients is impaired by the alcohol or drug's effect on body tissues, particularly the liver, pancreas and brain. Unintentionally losing 10 lb (4.5 kg) or more weight may be a sign of malnutrition. People who are malnourished may be skinny or bloated. Their skin is pale, thick, dry, and bruises easily. Rashes and changes in pigmentation are common. Hair is thin, tightly curled and pulls out easily. Joints ache and bones are soft and tender. The gums bleed, and tongue may be swollen or shriveled and cracked. Visual disturbances include night blindness, and increased sensitivity to light and glare. Anaemia, diarrhoea, disorientation, goiter, loss of reflexes, lack of coordination, muscle twitching, and scaling and cracking of lips and mouth are other symptoms (Merritt, 1978). All these signs and symptoms are correlated with those of many cancers (Pandey, 2007).

Role of Fruits and Vegetables in Malnutrition Cancer

Though many cancers can be controlled, the chronicity of the disease and expenses on supportive treatment are the valid reasons of not getting the cancer cured. Recent researches revolve round the urgency to evolve suitable chemotherapy consistent with new discoveries for the treatment of cancer with no toxic effect on normal cells. Toxicological aspects of drug warrant serious consideration without which the therapeutic concept of clinical efficacy will remain misconceived. Chemotherapy being a major treatment modality used for the control of advanced stages of malignancies and as a prophylactic against possible metastasis, exhibits severe toxicity on normal tissues (Madhuri, 2008; Pandey, 2007; Somkuwar, 2003).

If malnutrition results in cancer, it is very difficult to treat, but some improvements can be brought about by taking proper foods or nutrients and drugs (Pandey, 2007). The poor diets can cause diseases, or some diseases can be cured or controlled by particular diets. From ancient times India has endowed food with magical qualities. In the Vedic times, food was associated with divine attributes. This tradition was prevalent in other societies as well-ancient Egypt, for example. Ancient people discovered the healing powers of food perhaps initially by accident; later, these patterns became well established. Another episode in man's search for a cure is the story of scurvy. The scurvy is caused by vitamin C deficiency. Fresh herbs, oranges and other citrus fruits were found to be effective in preventing and controlling scurvy. These episodes in the history of man's battle against disease clearly illustrate the fact that food can cure certain types of diseases. Thus, there is no doubt that food can help to control several ailments (DNHE-2, 2003; Pandey, 2007). As long ago as 480 BC, Hippocrates recognized that several aspects of what we now call 'lifestyle' must come together to produce a healthy body. He said, 'Positive health requires a knowledge of man's primary constitution and the powers of various foods, both those natural to them and those resulting from human skill'. What Hippocrates called 'man's primary constitution', we today call 'genetics', and we can infer that foods 'resulting from human skills' can be equated with today's diet (Pandey and Madhuri, 2008a; Reddya, 2003).

To prevent energy deficiency, carbohydrate-rich vegetable sources like cereals, roots and tubers, fruits like as banana, sapota and mango, and fat-rich foods like nuts, oilseeds, vegetable oil, ghee and vanaspati should be consumed. If protein is lacking in the diet, vegetable foods such as pulses, nuts and oilseeds should be taken. If the vitamins and minerals are deficient in the diet, green leafy vegetables, nuts, pulses, etc. are to be eaten. In addition to these items, fiber also has a valuable role in the prevention of cancer of the colon. In cancer in which tissue wasting takes place, a high caloric-high protein diet is best. For this purpose, a suitable dietary guideline follows as: cereals- 400 g, pulses- 50 g, roots and tubers- 100 g, green leafy vegetable 200 g and other vegetables 200 g (DNHE-2, 2003).

It has been elucidated that many plant products exhibit potent anticancer activity against several cancer cell lines (Madhuri and Pandey,

2008; Polidori, 2003). Fruits and vegetables play an important role for the treatment and prevention of cancer. Consumption of large amount of fruits and vegetables can prevent the development of cancer. Many doctors recommend that people wishing to reduce their risk of cancer eat several pieces of fruits and several portions of vegetables every day. An inverse relationship has been suggested between the consumption of fruits and vegetables and the incidence of cancer in multiple organs including lung, larynx, mouth, pharynx, gastrointestinal tract and pancreas (Vecchia and Tawani, 1998). The intake of 400-600 g per day of vegetables and fruits can reduce the occurrence of many common forms of cancer, and diets rich in plant foods can also lower the risk of heart disease and many chronic diseases (Heber, 2004). Further, one-third of all cancer deaths in the United States could be avoided through dietary modification, which includes an abundant intake of fruits and vegetables. A report (Craig, 2006) of the WHO study on diet, nutrition and prevention of chronic diseases recommended that we daily consume at least 400 g of vegetables and fruits including at least 30 g of pulses, nuts and seeds. People who eat much quantity of vegetables have about one-half the risk of cancer and less mortality from cancer.

Epidemiological data as well *in vitro* studies strongly suggest that fruits and vegetables have strong protective effect against major diseases including cancer. The protective action of fruits and vegetables against cancer has been attributed to the presence of antioxidants, especially antioxidant vitamins (Madhuri, 2008). Fruits and vegetables contain several phytochemicals which possess strong antioxidant activities. Thus the fruits and vegetables prevent from cancer and other diseases by protecting cells from damage caused by 'free radicals' highly reactive oxygen compounds. Certain phytochemical antioxidants with anticancer activity include vitamins (A, C, E, K), carotenoids, terpenoids, polyphenols (ellagic acid, gallic acid, tannins), flavonoids (quercetin, anthocyanins, catechins, flavones, flavonones, isoflavones), enzymes (superoxide dismutase, catalase, glutathion peroxidase), minerals (Cu, Mn, Se, Zn), polysaccharides, saponins, lignins and xanthones (Gupta and Sharma, 2006; Madhuri, 2008; Madhuri and Pandey, 2009b and 2010; Pandey and Madhuri, 2008a; Ray and Hussan, 2002).

Fruits and vegetables contain compounds such as sulphoraphane that induces GSH transferase, there by helping detoxifying many types of carcinogens. Increased consumption of vegetables can increase the plasma antioxidant capacity and is associated with the lower risk of cancer (Rao *et al.*, 2004). The vegetables and fruits are most effective against those cancers that involve epithelial cells such as cancers of lung, esophagus, stomach, colon, pancreas and cervix. The protective effect of vegetables has also been observed for hormone related cancers (Vecchia and Tawani, 1998). The alcoholic extracts of dhaniya, haldi, karela and adrak have been tested for tumour growth inhibitory effects. Haldi and karela were most effective against metastatic prostate cancer cell lines (Rao *et al.*, 2004). Vegetables with the highest anticancer activity included lasun, soybean, pattagobhi, gajar; with a modest level of cancer-protective activity include piyaz, nibu, haldi, phoolgobhi, tamatar; with a low level of anticancer activity included khira (Madhuri and Pandey, 2008; Vecchia and Tawani, 1998).

There are many fruits and vegetable pigments such as flavonoids, carotenoids and anthocyanins which protect us from various diseases. Quercetin (a flavonoid) chiefly present in red and yellow piyaz and French bean possesses both anticarcinogenic activity. The carotenoids are powerful antioxidants that provide protection against oxidative damage, and stimulate immune function. Persons with high levels of serum carotenoids have a reduced risk of cancer. In addition, a variety of phenolic compounds (caffeic, ellagic and ferulic acids, sesamol, and vanillin) are present in fruits and vegetables. These exhibit antioxidant and anticancer activities. Adrak contains phenolic compounds (gingerol) which have antioxidant activity that is even greater than α-tocopherol (vitamin E). Compounds that stimulate the activity of glutathione S-transferase (GST) are considered as inhibitors of cancer. Substances that stimulate GST activity include the phthalides in celery seeds; sulphides in piyaz and lasun; dithiolthiones and isothiocyanates in sarson, rai, phoolgobhi and pattagobhi; limonoids in nibu; and curcumins in adrak and haldi (Heber, 2004; Madhuri, 2008). The phytochemicals such as allyl sulphides present in lasun and piyaz; glucarates in nibu, began and alu; phytates, lignans, isoflavones and saponins in soybean; isothiocyanates in sarson, phoolgobhi and pattagobhi; and flavonoids, carotenoids and terpenoids in other fruits and vegetables are reported to provide protection against cancer (Craig, 2006). These block various hormone actions and metabolic pathways associated with cancer. The isoflavones present in soybean is the contributing factor in the low incidence of breast, prostate, stomach, colon, rectum and lung cancers (Heber, 2004).

Nibu fruit due to its flavonoid, tangeretin and nobiletin contents can potentially inhibit the tumor cell growth and can activate the detoxifying cytochrome P-450 enzyme system. Limonoids present in nibu inhibit tumor formation by stimulating the enzyme GST. The limonene (a terpenoid) content of nibu also possesses anticancer activity. Nibu has been used for inhibiting the breast cancer cell proliferation, delaying of mammary tumorigenesis and curing the metastasis and leukemia cancers (Craig, 2006; Heber, 2004). Vegetables and fruits also contain a variety of isoprenoid compounds that exhibit antitumour/anticancer activities. These compounds include tocotrienols (related to tocopherols) and terpenoids (*e.g.*, limonene, geranoil, menthol and carvone). Overall, these compounds increase tumour latency and decrease tumour multiplicity (Heber, 2004; Vecchia and Tawani, 1998).

Conclusion

If the right food is not consumed in right amounts by a person it results in malnutrition, which may cause cancer. Hence, malnutrition is responsible for the development of "malnutrition cancer" as a consequence of which a huge mortality occurs every year in human beings, especially in children all over the world. If malnutrition results in cancer, it is very difficult to treat, but some improvements can be brought about by taking proper foods or nutrients and drugs. In spite of many allopathic treatments (*e.g.*, anticancer drugs or chemotherapy, radiation therapy, etc.), no relief or improvements occur. Hence, the individuals should be advised to take sufficient nutritious diets, including more amounts of various fruits and vegetables. The protective and curative actions of fruits and vegetables against cancer have been attributed on the ground of several scientific reports. It has been suggested that persons wishing to reduce their risk of cancer must eat several pieces of fruits and several portions of vegetables every day. Consumption of various fruits and vegetables play a significant role against malnutrition cancer.

References

Agrawala, S.K., Chatterjee, S. and Misra, S.K. (2001). Immune-potentiation activity of a polyherbal formulation "Immu-21" (Research Name). *Phytomedica* 2(1 and 2) : 1-22.

Copeland, E.M. 3rd, MacFadyen, B.V. Jr., Lanzotti, V.J. and Dudrick, S.J. (1975). Intravenous hyperalimentation as an adjunct to cancer chemotherapy. *Am. Surg.* 129(2) : 167-173.

Craig, W.J. (2006). Phytochemicals: Guardians of our health- website. DNHE-1 (2004). Nutrition-related disorders. In: *Nutrition for the Community*. Indira Gandhi National Open University (IGNOU), New Delhi, India.

DNHE-2 (2003). Dietary management of disease. In: *Public Health and Hygiene*. IGNOU, N. Delhi.

DNHE-3 (2004 a). Nutrition and health status of the community. In: *Nutrition and Health Education*. IGNOU, N. Delhi.

DNHE-3 (2004 b). Nutrition programmes. In: *Nutrition and Health Education*. IGNOU, N. Delhi.

Gottschlich, M.M. (2001). *The Science and Practice of Nutrition Support*. Kendall/Hunt Publishing, Dubuque.

Greenberg, D.S. (1978). Washington report: The Cancer Institute on the Griddlc. *N. Engl. J. Mcd.* 299(4) : 207-208.

Gupta, V.K. and S.K. Sharma (2006). Plants as natural antioxidants. *Nat. Prod. Rad.* 5(4): 326-334.

Heber, D. (2004). Vegetables, fruits and phytoestrogens in the prevention of diseases. *J. Postgrad. Med.* 50 : 145-149.

Kumar, V., Abbas, A.K. and Fausto, N. (2006). Neoplasia. In: *Pathologic Basis of Disease*, 7th Edn. Saunders, Elsevier India Pvt. Ltd., New Delhi, pp. 269-342.

Madhuri, S. (2008). *Studies on oestrogen induced uterine and ovarian carcinogenesis and effect of ProImmu in rat*. PhD thesis, RDVV, Jabalpur, MP, India.

Madhuri, S. and Pandey, Govind (2008). Some dietary agricultural plants with anticancer properties. *Pl. Arch.* 8(1) : 13-16.

Madhuri, S. and Pandey, Govind (2009 a). Environmental contaminants causing malnutrition and cancer. Abstract: 45 (Paper presented in National Seminar of Hawabagh Women's College held on 12th-13th January, 2009 at Jabalpur).

Madhuri, S. and Pandey, Govind (2009 b). Ethnomedicinal plants for prevention and treatment of tumours. *Int. J. Green Pharm.* 3(1) : 2-5.

Madhuri, S. and Pandey, Govind (2010). Effect of ProImmu, a herbal drug on estrogen caused uterine and ovarian cytotoxicity. *biomed* 5(1) : 57-62.

Merritt, R.J. (1978). *The ASPEN Nutrition Support Practice Manual.* ASPEN Silver Spring.

Pandey, Govind P. (1990). *Hepatogenic effect of some indigenous drugs on experimental liver damage.* PhD thesis, JNKVV, Jabalpur.

Pandey, Govind (2007). *Role of malnutrition in causing cancer and its safety measures.* Diploma in Nutrition and Health Education (DNHE) project report. IGNOU, N. Delhi.

Pandey, Govind and Madhuri, S. (2006 a). Medicinal plants: Better remedy for neoplasm. *Indian Drugs* 43(11) : 869-874.

Pandey, Govind and Madhuri, S. (2006 b). Autochthonous herbal products in the treatment of cancer. *Phytomedica* 7 : 99-104.

Pandey, Govind and Madhuri, S. (2008 a). Some anticancer agents from plant origin. *Pl. Arch.* 8(2) : 527-532.

Pandey, Govind and Madhuri, S. (2008 b). Agriculture plants used against livestock diseases. *Int. J. Plant Sci.* 3(2) : 671-674.

Pandey, Govind and Madhuri, S. (2010). Global scenario of radiation in cancer risk: A review. *Int. J. Pharmaceu. Biol. Arch.* 1(3) : 195-198.

Parkin, D.M., Bray, F., Ferlay, J. and Pisani, P. (2005). Global cancer statistics 2002. *Cancer J. Clin.* 55 : 74-108.

Polidori, M.C. (2003). Antioxidant micronutrients in the prevention of age-related diseases. *J. Postgrad. Med.* 49 : 229-235.

Pupim, L.B., Kent, P. and Hakim, R. (1999). The potential of intradialytic perenteral nutrition: A Review. *Minor Electrolyte Metab.* 25 : 317-323.

Rao, K.V.K., Schwartz, S.A., Nair, H.K., Aalinkeel, R., Mahajan, S. *et al.* (2004). Plant derived products as a source of cellular growth inhibitory pytochemicals on PC-3M, DU-145 and LNCaP prostate cancer cell lines. *Current Science* 87 : 1585-1588.

Ray, G. and S. A. Hussan (2002). Oxidant, antioxidant and carcinogenesis, *Indian J. Exp. Biol.* 40 : 1213-1232.

Reddya, L., Odhava, B., and Bhoolab, K.D. (2003). Natural products for cancer prevention: A global perspective. *Pharmacol. and Therap.* 99 : 1-13.

Rosangkima, G. and Prasad, S.B. (2004). Antitumour activity of some plants from Meghalaya and Mizoram against murine ascites Dolton's lymphoma. *Indian J. Exp. Biol.* 42 : 981-988.

Sivalokanathan, S., Ilayaraja, M. and Balasubramanium, M.P. (2005). Efficacy of *Terminalia arjuna* (Roxb.) on N-nitrosodiethylamine induced hepatocellular carcinoma in rats. *Indian J. Exp. Biol.* 43 : 264-267.

Somkuwar, A.P. (2003). *Studies on anticancer effects of Ocimum sanctum and Withania somnifera on experimentally induced cancer in mice.* PhD thesis, JNKVV, Jabalpur.

Vecchia, C.L. and Tawani, A. (1998). Fruits, vegetables and human cancer. *Eur. J. Cancer* 7 : 3-8.

आधुनिक जीवन शैली एवं कुपोषण से कैंसर: एक कयामत

गोविन्द पाण्डेय[1] और माधुरी शर्मा[2]

[1]पूर्व प्रोफेसर/प्रमुख वैज्ञानिक (फार्माकोलाजी), व प्रभारी अधिकारी, पशु माता महामारी, जबलपुर संभाग, जबलपुर, म.प्र. शासन

E-mail : drgovindpandey@rediffmail.com

[2]सी.एस.आई.आर. रिसर्च फेलो, प्राणि विज्ञान व बॉयोटेक्नॉलाजी विभाग, शासकीय आदर्श विज्ञान महाविद्यालय, जबलपुर, म.प्र.

संपूर्ण विश्व में प्रति वर्ष 1.1 करोड़ लोग कैंसर से पीड़ित होते हैं । भारत वर्ष में प्रतिवर्ष कैंसर के 7 लाख मरीज बढ़ रहे हैं । पूरी दुनिया में प्रति वर्ष 70 लाख लोग इस रोग के कारण काल-कवलित हो जाते हैं । भारत में हर साल 3.5 लाख मनुष्य कैंसर की वजह से मृत्यु की गोद में समा जाते हैं । इनमें सबसे अधिक फेफड़े के कैंसर का निशाना बनते हैं । भारत सरकार के स्वास्थ मंत्रालय ने कैंसर से संबंधित जो आँकड़े जारी किए हैं, वह किसी का भी दिल दहलाने के लिए काफ़ी हैं । सन् 2025 तक भारत में कैंसर के रोगी 500 गुना बढ़ जाएँगे । इनमें से 280 प्रतिशत लोगों को होने वाले कैंसर के कारण कुछ भी हों, परंतु 220 प्रतिशत को यह रोग तंबाकू के कारण होगा । इसके साथ ही एक भयावह सच यह भी है कि तंबाकू के उपयोग में हुई कमी के कारण कैंसर फैलने की गति तो घटी, किंतु दुर्भाग्य से उसकी जगह शहरों की विषाक्त हवा और पर्यावरण ने ले ली । शहरी भारत में फैला प्रदूषण कैंसर का एक बड़ा कारण है, विशेषकर मुंबई में । हमारी जीवनशैली में हुए बदलाव और कुपोषण कैंसर का बहुत बड़ा कारण बन रहे हैं । प्रतिवर्ष कैंसर के 10 लाख मामले सामने आते हैं, उनमें 65 से 80 प्रतिशत मरीजों के बारे में कहा जा सकता है कि उन्हें आधुनिक जीवनशैली व कुपोषण

[Paper published in: Science India (Hindi), 2010; 3(3): 24-26].

के कारण कैंसर हुआ । बहुत से लोग शारीरिक व्यायाम बिल्कुल नियम से करते हैं । इसके बावजूद वे कैंसर से ग्रसित हो जाते हैं तो इसमें बड़ा हाथ हमारी अनियमित दिनचर्या का है । पुरूषों को मुख्यतः ब्रेन ट्यूमर, फेफड़े या गले के कैंसर अधिक होते हैं । इसका एक बड़ा कारण पुरूषों का धूम्रपान करना या तंबाकू सेवन करना है। पुरूषों को पौरूष ग्रंथि का कैंसर भी अधिक होता है । महिलाओं को होने वाले कैंसर में सबसे बड़ा प्रतिशत स्तन (ब्रेस्ट) कैंसर का है तथा इसके बाद गर्भाशय द्धबच्चेदानीत्रः व गर्भाशय की ग्रीवा (सर्विक्स) कैंसरों का स्थान आता है । विशेषज्ञों के अनुसार ये कैंसर अक्सर देरी से विवाह व संतान होने के कारण होते हैं । ज़ाहिर है, विवाह में देरी का कारण हमारी जीवनशैली में आया बदलाव ही है । खानपान में लापरवाही से कई तरह के कैंसर का सामना करना पड़ता है । अगर हम अपनी जीवनशैली में सुधार कर लें तो देश के 50 प्रतिशत लोगों को कैंसर से बचा सकते हैं । पहले तो कैंसर उम्रदराज़ लोगों को ही होता था, पर अब यह युवाओं को भी होने लगा है। बढ़ती जनसंख्या, लापरवाही और गरीबी भारत में कैंसर फैलने के प्रमुख कारण हैं । तथापि कैंसर से लड़ने के लिए नई-नई दवाएँ और अत्याधुनिक तकनीक हमारे यहाँ उपलब्ध हैं । हमारे वैज्ञानिक, चिकित्सक और सरकार इस मामलें में बहुत गंभीर हैं। सुखद बात यह है कि कैंसर से निपटने का हर तरीका हमारे यहाँ उपलब्ध है । भारी चीर-फाड़ की जगह छोटे ऑपरेशन होने लगे हैं । हम पूरे अंग को अस्त-व्यस्त करने के बजाय केवल ट्यूमर को ही निशाना बनाते हैं । अलबत्ता छोटे कैंसर चिकित्सा केन्द्रों को अधिक विकसित करने की आवश्यकता है ।

भूमिका

कैंसर कयामत ढा रहा है । प्रतिदिन भारत और संपूर्ण विश्व में अनगिनत लोग इसका शिकार बन रहे हैं । हमने विकास के नाम पर ख़ुद अपने लिए ज़हर बो लिया है, जिस पर कैंसर जैसे रोग फल-फूल रहे हैं । हमारी जीवन पद्धति में कैंसर के लिए छिपा है आमंत्रण । वक़्त है, अपनी सच्चाई जानने और इस कठिनाई से जूझने का । क्योंकि भारत सरकार के स्वास्थ्य मंत्रालय ने कैंसर से संबंधित जो आंकड़े जारी किए हैं, वह किसी का भी दिल दहलाने के लिए काफ़ी हैं । इन आंकड़ों के अनुसार सन् 2025 तक भारत में कैंसर के रोगी 500 गुना बढ़ जाएंगे । इनमें से 280 प्रतिशत लोगों को होने वाले कैंसर के कारण कुछ भी हों, परंतु 220 प्रतिशत को यह रोग तंबाकू के कारण होगा ।

इसके साथ ही एक भयावह सच यह भी है कि तंबाकू के उपयोग में हुई कमी के कारण कैंसर फैलने की गति तो घटी, किंतु दुर्भाग्य से उसकी जगह शहरों की विषाक्त हवा और पर्यावरण ने ले ली । शहरी भारत में फैला प्रदूषण कैंसर का एक बड़ा कारण है, विशेषकर मुंबई व अन्य महानगरों के लिए तो यह बिल्कुल सच है । वहां पर हरियाली की भयंकर कमी है । कारों की तादाद बढ़ती जा रही है तथा औद्योगिक कचरा शहर की हवा और पानी में लगातार घुलता जा रहा है । वर्तमान में हमारी जीवनशैली

में हुए बदलाव और कुपोषण कैंसर का बहुत बड़ा कारण बन रहे हैं । प्रतिवर्ष कैंसर के 10 लाख मामले सामने आते हैं, उनमें 65 से 80 प्रतिशत मरीजों के बारे में कहा जा सकता है कि उन्हें आधुनिक जीवन शैली व कुपोषण के कारण कैंसर हुआ ।

आधुनिक जीवन शैली एवं कुपोषण से कैंसर

इस समय भारत में कैंसर के उपचार हेतु विश्व की सर्वोत्तम तकनीक और इलाज की सुविधायें मौजूद हैं, इसके बावजूद ख़राब भोजन और बेढंगी जीवनशैली के कारण कैंसर लगातार फैलता जा रहा है । हालत यह है कि एहतियात करते हुए भी लोग कैंसर की चपेट में आ ही जाते हैं । बहुत से लोग शारीरिक व्यायाम बिल्कुल नियम से करते हैं । इसके बावजूद वे कैंसर से ग्रसित हो जाते हैं तो इसमें बड़ा हाथ हमारी अनियमित दिनचर्या का है ।

पुरूषों को मुख्यतः ब्रेन (मस्तिष्क) ट्यूमर, फेफड़े या गले के कैंसर अधिक होते हैं। इसका एक बड़ा कारण पुरूषों का धूम्रपान करना या तंबाकू सेवन करना है । पुरूषों को पौरूष ग्रंथि का कैंसर भी अधिक होता है । महिलाओं को होने वाले कैंसर में सबसे बड़ा प्रतिशत स्तन (ब्रेस्ट) कैंसर का है तथा इसके बाद गर्भाशय (बच्चेदानी) व गर्भाशय की ग्रीवा (सर्विक्स) कैंसरों का स्थान आता है । विशेषज्ञों के अनुसार ये कैंसर अक़्सर देरी से विवाह व संतान होने के कारण होते हैं । ज़ाहिर है, विवाह में देरी का कारण हमारी जीवनशैली में आया बदलाव ही है । जानकारी के अनुसार 25 साल की उम्र में पहुंचने से पहले मां बनने वाली महिलाओं को छाती के कैंसर की आशंका आधी रह जाती है, बजाय उन महिलाओं के जिन्हें देर से संतान प्राप्ति होती है । विवाह के बाद ऐसी महिलाएं जब अपना परिवार बढ़ाना शुरू करती हैं तो वह स्तन (छाती) कैंसर का आसान निशाना होती हैं । हालांकि हमारे यहां की स्थिति अमेरिका से बेहतर है । देर से मां बनने वाली महिलाओं में छाती के कैंसर की आशंका अमेरिका में अधिक है । वहां ऐसी 9 महिलाओं में से 1 को यह बीमारी होती है, जबकि हमारे यहां 20 में से 1 को होती है । किंतु वह हमारी चिंता बढ़ा देते हैं कि हमारे यहां अब इसका प्रतिशत बढ़ता जा रहा हैं । इस मामले में हमारी लापरवाही भी एक बड़ा कारण है । कुछ साल पहले जसलोक अस्पताल द्वारा कराए गये एक सर्वे से पता चला है कि मुंबई की शिक्षित महिलाओं में से केवल 1 प्रतिशत ने इसकी जांच कराई कि उन्हें छाती का कैंसर है या नहीं ।

हमारी जीवन शैली के बदलाव में रेडियो, दूरदर्शन (टी वी), कम्प्यूटर, मोबाइल फोन तथा घरेलू उपयोगी विद्युत सामग्री भी सम्मिलित हैं, जिनके कारण कैंसर की संभावना अधिक होती है । एफ एम रेडियो का अधिक उपयोग करने से त्वचा का कैंसर होता है । वयस्क महिलाओं में स्तन कैंसर कम्प्यूटर का उपयोग (प्रतिदिन 3 घंटे से अधिक) तथा मोबाइल फोन, टी वी सेट एवं घरेलू विद्युत उपकरणों के अत्यधिक करने से होता है ।

खानपान में लापरवाही से कई तरह के कैंसर का सामना करना पड़ता है । हम ज़्यादातर जंक या डिब्बा बंद खानों का इस्तेमाल करने लगे हैं । इनमें तरह-तरह के रसायन मिले होते हैं । पहले इस तरह का भोजन हम नहीं करते थे । हमारे पास अब समय ही नहीं है अच्छी तरह से घर में पका हुआ भोजन खाने का । परंतु इससे बचा जा सकता है । वस्तुतः पूरे विश्व में होने वाले कैंसर का 35 प्रतिशत कैंसर अनियमित या ग़लत भोजन के कारण होता है । बड़ी आँत का 80 प्रतिशत कैंसर कुपोषण के कारण होता है । इस तरह कुपोषण कैंसर सहित अनेक बीमारियों को मनुष्यों एवं पशुओं में जन्म देता है । कुपोषण के कारण प्रतिवर्ष लगभग 3 लाख 5 साल से कम उम्र के बच्चे विकासशील देशों में मर जाते हैं।

द नेशनल कैंसर रजिस्ट्री के अनुसार इस तरह का कैंसर लगातार बढ़ रहा है, जिसमें अगर हम एक सरसरी नज़र से ही देखें तो 50 प्रतिशत कैंसर जीवन शैली में आए बदलाव के कारण होता है। अगर हम अपनी जीवन शैली में सुधार कर लें तो देश के 50 प्रतिशत लोगों को कैंसर से बचा सकते हैं । पहले तो कैंसर उम्रदराज़ लोगों को ही होता था, परंतु अब यह युवाओं को भी होने लगा है । इसके मुख्य कारण जीवन शैली और वातावरण में आ रहे बदलाव ही हैं । बढ़ती जनसंख्या, लापरवाही और गरीबी भारत में कैंसर फैलने के मुख्य कारण हैं ।

हमारे यहाँ कैंसर से लड़ने के लिए नई-नई दवाएँ और अत्याधुनिक तकनीक हमारे यहाँ उपलब्ध हैं । हमारे वैज्ञानिक, चिकित्सक व सरकार इस मामलें में बहुत गंभीर हैं। सरकार ने दिल की बीमारियों के बाद कैंसर को ही प्राथमिकता दे रखी है । इस घातक रोग का सामना करने के लिए हमारे पास धन है । इसके लिए नेशनल कैंसर कंट्रोल प्रोग्राम (एनसीसीपी) के बजट में हाल ही अत्यधिक वृद्धि की गई है । नौवीं पंचवर्षीय योजना (1997-2002) में संस्था का बजट मात्र 85 करोड़ रूपये था । दसवीं पंचवर्षीय योजना (2002-07) में इसका बजट बढ़ा कर 260-285 करोड़ रूपए कर दिया गया ।

हमारे पास विश्वस्तरीय सुविधायें उपलब्ध हैं । बड़े ऑपरेशन से लेकर कीमोथेरेपी और रेडियोथेरेपी तक हर उपाय हमारे बड़े केन्द्रों में उपलब्ध हैं । अलबत्ता छोटे कैंसर चिकित्सा केन्द्रों को अधिक विकसित करने की आवश्यकता है । अब कैंसर का उपचार और अधिक बारीक़ी से होने लगा है । भारी चीर-फाड़ की जगह छोटे ऑपरेशन होने लगे हैं । हम पूरे अंग को अस्त-व्यस्त करने के बजाय केवल ट्यूमर को ही निशाना बनाते हैं। उदाहरणार्थ, स्तन कैंसर में अब पूरी छाती निकालने की जरूरत नहीं पड़ती, जैसा कि पहले होता था । यद्यपि भारत में औसत उम्र बढ़ने के साथ-साथ कैंसर के रोगी भी बढ़ते जाएंगे । किंतु हम रोगी को रोगमुक्त करने के साथ ही उसकी शेष ज़िंदगी को भी आरामदेह बनाने में विश्वास रखते हैं ।

सर्वाधिक होने वाले कैंसर

भारत में दो तरह के कैंसर बहुत तेज़ी से फैल रहे हैं । एक है पौरूष (प्रोस्टेट) ग्रंथि का और दूसरा पित्ताशय का कैंसर । पौरूष ग्रंथि वाला कैंसर फैलने का मुख्य कारण

पुरूषों के यौन संबंधी हार्मोन्स में परिवर्तन आना है, जबकि पित्ताशय का कैंसर खाने में ख़राब तेल का उपयोग करने से होता है । एक अनुमान के अनुसार विकसित देशों में सन् 2020 तक जिस कैंसर के कारण सबसे ज़्यादा जानें जाएंगी, वह है पौरूष ग्रंथि का कैंसर । इसी तरह इस अवधि के दौरान विकासशील देशों के लिए सबसे बड़ा खतरा होगा फेफड़े का कैंसर ।

कैंसर संबंधी कुछ महत्वपूर्ण आंकड़े

- ☆ संपूर्ण विश्व में प्रतिवर्ष 1.1 करोड़ कैंसर रोगी सामने आते हैं ।
- ☆ भारत में प्रतिवर्ष कैंसर के 7 लाख रोगी बढ़ रहे हैं ।
- ☆ पूरे विश्व में प्रतिवर्ष 70 लाख लोग इस रोग के कारण मौत का शिकार होते हैं ।
- ☆ हमारे देश में प्रतिवर्ष 3.5 लाख लोग कैंसर के कारण काल के गाल में समा जाते हैं । इनमें सबसे अधिक लोग फेफड़े के कैंसर का निशाना बनते हैं ।
- ☆ हमारे देश में अभी लगभग 500 कैंसर विशेषज्ञ हैं, जबकि अमेरिका में इनकी संख्या लगभग 2 लाख है ।

Global Scenario of Radiation in Cancer Risk: A Review

Govind Pandey[1] *, *Madhuri Sharma*[2]

[1]*Rinder Pest (Veterinary/Animal Husbandry Department, Govt. of MP), Jabalpur Division, Jabalpur, M.P., India*

[2]*Department of Zoology and Biotechnology, Govt. Model Science College, Jabalpur, MP, India*

ABSTRACT

Emission of energy from any source is known as 'radiation'. High frequency radiation, *e.g.*, ionizing radiation and ultraviolet (UV) radiation may cause genetic damage, leading to cancer. The ionizing radiations include X-rays, γ-rays, cosmic rays and particles given off by radioactive materials, such as α and β particles, protons and neutrons. Radiant energy (UV rays of sunlight, ionizing electromagnetic or particulate radiation) induced the neoplasm (cancer) in both human beings and experimental animals. The ionized molecules are unstable and quickly undergo chemical changes, thereby forming the 'free radicals' that can damage the DNA molecule or other molecules around it. Hence, the ionizing radiations can cause mutation (change) in a cell's DNA, resulting into cancer. Thyroid gland and bone marrow are most sensitive to radiations; while kidney, bladder and ovary are least affected. Some forms of leukemia (bone marrow cancer) appear to be the most common

---

Received: 8 May 2010; Accepted 22 July 2010.

[Paper published in: International Journal of Pharmaceutical and Biological Archives, 2010; 1(3): 195-198].

radiation-induced cancers. The atomic blasts in Japan showed that high-dose radiation increases the risk of developing several cancers, including leukemia. Breast cancer can be developed in elderly women by a much more extensive use of personal computers (more than 3 hours a day), mobile telephones, TV sets and other household electrical appliances. Children appear to be twice as sensitive as adults to the leukemia-causing effects of radiation, and unborn children exposed to radiation in the uterus are even more sensitive. The breast cancer risk is more than twice as high as normal; however, the risk of developing lung cancer is 50 per cent higher and the risk for multiple myeloma is more than twice as high as in the general population. The radiation therapy at high doses can cause DNA mutations in cells that survive the radiation, which may lead to a second primary cancer. Of all the types of non-ionizing radiations, only UV rays are cancer-causing agents. The most skin cancers are a direct result of sunlight exposure as sun is the major source of UV radiation.

Keywords: *Cancer, Radiation, UV rays, X-rays, γ-rays*

Introduction

Cancer is an abnormal growth and proliferation of cells. It is a frightful disease which is uncontrollable and incurable, and may occur at any time at any age in any part of the body. Cancer is caused by a complex, poorly understood interplay of genetic and several environmental factors. It represents the largest cause of mortality in the world and several environmental factors. It represents the largest cause of mortality in the world and claims over 6 millions.[1–3] Many studies showed that several environmental factors, including air, water and industrial pollutants, environmental chemicals and radiations, etc. may cause various types of cancer.[4–6]

Emission of energy from any source is known as 'radiation'. Presently, there is hardly any aspect of human welfare in which nuclear radiation does not play an important role. The use of radiation has become an integral part of modern life. Radiation is used for scientific purposes, medical reasons and for power and energy generation. It has been stated that the high frequency radiations, *e.g.*, ionizing and UV radiations may cause genetic damage, leading to cancer.[4] Radiant energy, whether in the form of the UV rays of sunlight or as ionizing electromagnetic and particulate radiations, induced the neoplasm (cancer) in both human beings and experimental animals. The ionizing radiations include X-rays, γ-rays, cosmic rays and particles given off by radioactive materials, *e.g.*, α and β particles, protons and neutrons. The ionizing radiations consist of high-energy waves that are able to penetrate cells and can cause ionization. It has been narrated that the ionized molecules are unstable and quickly undergo chemical changes, thereby forming the free radicals that can damage the molecule or other molecules around it. One type of molecule that is sensitive to ionizing radiation is the DNA. Thus, the ionizing radiations can lead to a mutation (change) in a cell's DNA, which could contribute to cancer, or to damage/death of the cell. The amount of damage is related to the dose of radiation received by the cell. Organs differ in their sensitivity to the effects of radiation. The thyroid gland and bone marrow are most sensitive to radiations; while the kidney, bladder, and ovary seem to be least affected. Some forms of leukemia (the cancer of bone marrow) appear to be the most common radiation-induced cancers. Ionizing radiation is an effective way to treat certain kinds of cancer.[7] The cancer of thyroid gland follows closely but only in the young persons; while in the intermediate category are the cancers of breast, lung and salivary gland.[4]

Over all, the present article has been put forth to elucidate the various kinds of radiation which can develop different cancers in humans and animals. This knowledge would be definitely fruitful to the oncologists, biologists, pharmacologists and pharmaceutical scientists.

Effects of Radiation

Normally, the health is affected on the basis of amount and duration of radiation exposure. The health can be reduced or death may occur due to radiation exposure as mentioned below:[8]

1. Stochastic Health Effects

The stochastic effects are associated with long-term, low-level (chronic) exposure to radiation. 'Stochastic' refers to the likelihood that something will happen. Increased levels of exposure make these health effects more likely to occur, but do not influence the type or severity of the effect. Cancer is considered as the primary health effect from radiation exposure. The radiations can cause changes in DNA, which are called mutations.

2. Non-stochastic Health Effects

These effects appear due to high levels of radiation exposure and become more severe as the exposure increases. The short-term, high-level exposure is known as 'acute' exposure. Many non-cancerous health effects of radiations are non-stochastic.

3. Other Long-term Health Effects

Other than cancer, the most prominent long-term health effects are mutations. The mutations can be teratogenic or genetic. The genetic mutations are those which can be passed from parent to child (offspring). The teratogenic mutations are caused by exposure of the fetus in uterus, and affect only the individual who was exposed. The teratogenic mutations can be smaller head or brain size, poorly formed eyes, abnormally slow growth and mental retardation.

Various Radiations Causing Cancer

The childhood leukemia was early connected to power-frequent magnetic fields of radiation already in the pioneering work by Wertheimer and Leeper[9], and more recently Scandinavian scientists have identified an increased risk of malignant brain tumors, *e.g.*, astrocytoma and meningioma.[10–11] Furthermore, a clear association between adult cancers and FM radio broadcasting radiation has been noticed, both in time and location.[12] The breast cancer developed in elderly women by a much more extensive use of personal computers (more than 3 hours a day), mobile telephones, TV sets and other household electrical appliances. Among the elderly women who developed breast cancer in the first time frame, 20 per cent were regularly exposed to power-frequent fields. But in the more modern period 51 per cent were so exposed, mainly through the use of personal computers. A significant influence of electromagnetic fields on the formation of epithelial mammary tumors was found.[13] It has already been pointed out[12] that there is a strong association between the body-resonant non-ionizing radiation (FM radio, 100 MHz) and the existence of malignant melanoma of skin. Since this frequency range has a penetration depth of about 10 cm into the human body there is a suspicion that resonant currents may affect the immune defense system also when it comes to beating cancer cells in lungs. Due to that, it is well motivated to study in detail how the presence and rate of lung cancer have changed in Sweden, UK and other countries as this new environmental factor was added.

No safe level or threshold of ionizing radiation has been said. Even the exposure to background radiation causes some cancers. Additional exposures cause additional risks. The hazards of exposure to some kinds of radiation were recognized shortly after the discovery of x-ray in 1895. The skin reactions were observed in many people working with early x-ray generators, and by 1902 the first radiation-caused cancer was reported in a skin sore. Within a few years, a large number of such skin cancers had been observed. The first report of leukemia in radiation workers appeared in 1911. Marie Curie (discoverer of radium) and her daughter are believed to have died of radiation-caused leukemia. Since then, many studies have confirmed the cancer-causing effects of some types of radiation.[7] The sun is the major source of UV radiation. There is ample evidence from epidemiologic studies that UV rays derived from the sun induce an increased incidence of basal cell and squamous cell carcinomas (the most common types of skin cancer), and possibly malignant melanoma of skin.[14] It has also been mentioned that the UV rays have a number of effects on cells, inhibition of cell division, inactivation of enzymes, induction of mutation, and in sufficient dosage, death of cells.[4] Further, the relationship between UV radiation and cancer of the face, lower lip and other exposed areas (especially in fair-skinned individuals) is considered to be causal for a long period of time, particularly in people with outdoor occupation or rural residents.[15]

Both electromagnetic (*viz.*, X-rays and γ-rays) and particulate (*viz.*, α and β particles, protons, and neutrons) radiations are carcinogenic. The evidence is so voluminous that a few examples suffice. Many pioneers of X-rays developed the skin cancers. Miners of radioactive elements in central Europe and Rocky Mountain region of the USA have a tenfold increased incidence of lung cancers; most telling is the follow-up of survivors of the atomic bombs in Hiroshima and Nagasaki. Initially, there was increased incidence of leukemias, mainly acute and chronic myelocytic leukemia after an average latent period of about 7 years. Subsequently, the incidence of many solid tumors of breast, colon, thyroid and lung increased with longer latent periods. In humans, there is a hierarchy of vulnerability of different tissues to radiation induced cancers. Most frequent are the leukemias, except chronic lymphocytic leukemia. An increased risk of breast cancer has become apparent decades later among women exposed during childhood to the atomic bomb.[16] A marked increase in thyroid cancer incidence was also seen in areas exposed to the fallout from the nuclear power plant accident in Chernobyl in 1986.[17]

Therapy by ionizing radiation is an effective way to treat certain types of cancer. During radiation therapy, high doses of ionizing radiation are directed at the cancer, resulting in the death of the cancer cells. However, this can lead to DNA mutations in cells that survive the radiation, which can eventually lead to another cancer, called a second primary cancer. An increase in second primary cancers in the area being irradiated has been observed in patients with several types of cancer following radiation therapy and/or chemotherapy. Treatment for Hodgkin disease (a type of lymphoma) often delivers lower radiation doses to many areas of the body. These treated areas include large amounts of normal tissue. Patients with Hodgkin disease who are treated with radiation therapy are at an increased risk for developing second primary tumors. When considering radiation exposure from radiation therapy treatment, the benefits generally outweigh the risks. However, some combinations of radiation therapy and chemotherapy are more risky than others. Children treated with radiation therapy for hereditary retinoblastoma, a malignant eye tumor, are at an increased risk for developing a type of bone cancer called osteosarcoma. Similarly, people who have nevoid basal cell carcinoma syndrome, a type of skin cancer, are at high risk for development of basal cell cancers in irradiated areas.[7]

References

1. Brunton LL, editor. Goodman and Gilman's The Pharmacological Basis of Therapeutics, 11th ed. New York: McGraw-Hill Co.; 2006, pp. 1315-1403.
2. Madhuri S. Studies on oestrogen induced uterine and ovarian carcinogenesis and effect of ProImmu in rats. PhD thesis, Rani Durgavati Vishwa Vidyalaya, Jabalpur, MP, India; 2008.
3. Pandey Govind and Madhuri, S. Medicinal plants: Better remedy for neoplasm. Indian Drugs 2006; 43(11):869-874.
4. Kumar V, Abbas AK, Fausto N. Pathologic Basis of Disease, 7th ed. New Delhi: Saunders, Elsevier India Pvt. Ltd.; 2006, pp. 269-342.
5. Mukherjee S, Bidhan CK, Ray S, Ray A. Environmental contaminants in pathogenesis of breast cancer. Indian J Exp Biol 2006; 44:597-617.

6. Pandey Govind. Role of malnutrition in causing cancer and its safety measures. Diploma in Nutrition and Health Education project report. Indira Gandhi National Open University, New Delhi; 2007.
7. www.cancer.org/docroot/PED/content/PED_1_3X_Radiation_Exposer_and_Cancer.asp.www.icucec.org/files/radiationbertell.pdf.
8. Bhatia A.L. Effect of radiation and gene mediated radiation protection with special reference to melatonin. Proceedings (Part II, Abstracts-Animal, Veterinary and Fishery Sciences) of 95th Indian Science Congress of the ISCA, Visakhapatnam, AP, 3-7 January, 2008; pp. 1-51.
9. Wertheimer N, Leeper E. Electrical wiring configurations and childhood cancer. Am J Epidemiol 1979; 109:273-284.
10. Hardell L, Carlberg M, Mild KH. Case-control study on cellular and cordless telephones and the risk for acoustic neuroma or meningioma in patients diagnosed 2000-2003. Neuroepidemiology 2005; 25:120-128.
11. Lonn S, Ahlbom A, Hall P, Feychting M. Mobile phone use and the risk of acoustic neuroma. Epidemiology 2004; 15:653-659.
12. Hallberg A, Johansson O. FM broadcasting exposure time and malignant melanoma incidence. Electromag Biol Med 2005; 24:1-8.
13. Beniashvili D, Avinoachm I, Baasov D, Zusman I. The role of household electromagnetic fields in the development of mammary tumors in women: Clinical case-record observations. Med Sci Monit 2005; 11:10-13.
14. Cleaver JE, Crowley E. UV damage, DNA repair and skin cancinogenesis. Front Biosci 2002; 7:1024.
15. Sen R. Oral cancer: The global scenario. Proceedings (Part II, Abstracts-Medical Science) of 95th Indian Science Congress of the ISCA, Visakhapatnam, AP, 3-7 January, 2008; pp. 1-26.
16. Preston *et al.*, Radiation effects on breast cancer risk: a pooled analysis of eight cohorts. Radiat Res 2002; 158:220.
17. Williams D. Chernobyl, 15 years later, correlation of clinical, epidemiological and molecular outcomes. Ann Endocrinol 2003; 64:72.

Some Plants as Source of Anticancer Drugs

***Govind Pandey*[1] *and Madhuri Sharma*[2]**

[1]*Ex-Professor and Head of Pharmacology (Pharmacy); Presently Officer-In-Charge of Rinder Pest (Animal Husbandry/Veterinary Department, Govt. of MP), Jabalpur Division, Jabalpur, M.P.*

E-mail: drgovindpandey@rediffmail.com, drgovindpandey@yahoo.co.in

[2]*Department of Zoology, Govt. MH College of Home Science and Science for Women, Jabalpur*

Cancer is the second leading cause of death in developed and developing countries. Natural (plant) products are the most important part of our life, many of which are consumed daily with the diet. They provide significant protection against various cancers and many other diseases. According to WHO estimates, more than 80 per cent people in developing countries depend on traditional medicine for their primary health needs. Consumption of large amount of vegetables and fruits can prevent the development of cancer. Several plant products are available as chemoprotective agents against various types of cancer. These are present in fruits, vegetables, plant extracts, herbs, microbes and marine organisms. A host of natural product constituents could be responsible for the protective effect against cancer. Although the mechanism of the protective effect is unclear, nevertheless, the consumption of natural

(Abstract paper published in: Souvenir of National Seminar organized by Hitkarini Mahila Mahavidyalaya, Jabalpur, MP held on November 23-24, 2010; pp. 38-39).

products lowers the incidence of cancer. A major group of these products are the powerful antioxidants, others are phenolic in nature and the remainder includes reactive groups. The antioxidant medicinal plants and their products prevent from cancer and other diseases by protecting the cells from damage. Thus, consuming a diet rich in antioxidant fruits, vegetables, herbs, etc. will certainly provide health-protective effects. Microbes and marine organisms also play an important role in the prevention and treatment of cancer. This paper emphasizes many anticancer drugs obtained from various sources of plant kingdom. Some important anticancer plants include *Allium cepa* (Piyaz), *Azadirachta indica* (Neem), *Camellia sinensis* (Tea), *Emblica officinalis* (Amla), *Fragaria vesca* (Strawberries), *Ocimum sanctum* (Tulsi), *Ricinus communis* (Arand), *Solanum* spp., *Swertia chirata* (Chirayita), *Triticum aestivum* (Wheat) and *Withania somnifera* (Ashwagandha). However, *Streptomyces* spp. is the main microbial source for different anticancer agents; whilst, some marine organism-derived anticancer agents are aplidin, bryostatin, citarabine, cryptophycin, discodermolide, dolastatin, ecteinascidin and halicondrin.

Antioxidant Vegetables Act against Cancer and Other Diseases

Govind Pandey

Ex-Professor and Head of Pharmacology (Pharmacy), presently, Officer-In-Charge of Rinder Pest (Veterinary/AH Department, Govt. of MP), Jabalpur Division, Jabalpur, M.P., India

E-mail: drgovindpandey@rediffmail.com, drgovindpandey@yahoo.co.in

ABSTRACT

More than 80 per cent people in developing countries depend on traditional medicine or plants and their products for their primary health needs. Indigenous medicinal plants cure different diseases, including cancer without causing any toxicity. Vegetables prevent human from several severe and chronic diseases. We must consume daily at least 400 g of vegetables and fruits including at least 30 g of pulses, nuts and seeds. People who eat much quantity of vegetables have about one-half the risk of cancer and less mortality from cancer. Antioxidant vegetables prevent the human from many severe and chronic diseases, including cancer, cardiovascular disease, diabetes, hypertension, leprosy, rheumatism, epilepsy, liver and urinary disorders, stroke, inflammation, paralysis, asthma, diarrhoea, etc. The vegetables contain several phytochemicals possessing antioxidant activity. The antioxidants protect the cells from damage caused by 'free radicals' the highly reactive oxygen compounds. Major groups of the vegetable phytochemicals which shows antioxidant

---

[Paper published in: International Journal of Pharmaceutical Studies and Research, 2011; 2(1): 32-38].

activity include vitamins (A, C, E, K), carotenoids, terpenoids, flavonoids, polyphenols, saponins, enzymes and minerals. All these compounds prevent from various cancers and other diseases. Hence, the present article explore out some antioxidant vegetables which are commonly taken in the diet, and act against cancer and other diseases.

Keywords: *Antioxidants, Cancer and other diseases, Indigenous medicinal plants, Phytochemicals, Vegetables.*

Introduction

Plants and their products have been used for treating various diseases of humans as well as animals since times immemorial. According to WHO estimates, more than 80 per cent people in developing countries depend on traditional medicine for their primary health needs. Ethnomedicinal survey exhibits a number of valuable data on indigenous medicinal plants, which are being used against various diseases such as cancer, diseases of heart, liver, lung, stomach, intestine, spleen and skin, diabetes, diarrhea, hypertension, rheumatism, nervous disorders, stroke, inflammation, fever, cough, cold and ulcer. Thus the plants not only maintain the health and vitality of individuals, but also cure different diseases, including cancer without causing any toxicity. A recent survey shows that more than 60 per cent of cancer patients use medicinal plants at some point in their therapy (1-4).

Cancer is a very severe disease, which kills annually about 3500 per million population allover the world. Several chemopreventive agents are used to treat cancers, but they cause toxicity that prevents their usage. Although more than 1500 anticancer drugs are in active development with over 500 of the drugs under clinical trials, there is an urgency to investigate much effective drugs with no toxicity. In India, only limited research works have been done to cure cancer by herbal extracts which are mainly confined to adrak, tamatar and neem (5-6). As one of the major plant resources (natural products), the vegetables prevent humans from several severe and chronic diseases. Consumption of fruits and vegetables is widely accepted as lowering the risk of most common cancers. Many doctors recommend that people wishing to reduce their risk of cancer eat several pieces of fruits and several portions of vegetables every day. An inverse relationship has been suggested between the consumption of fruits and vegetables and the incidence of cancer in multiple organs, including lung, larynx, mouth, pharynx, gastrointestinal tract and pancreas (1,7-8). A report of the WHO study on diet, nutrition and prevention of chronic diseases recommended that we daily consume at least 400 g of vegetables and fruits including at least 30 g of pulses, nuts and seeds. People who eat much quantity of vegetables have about one-half the risk of cancer and less mortality from cancer (9). The intake of 400-600 g/day of vegetables and fruits can reduce the occurrence of many common forms of cancers, and diets rich in plant foods can also lower the risk of heart disease and many chronic diseases (10). The author further stated that one-third of all cancer deaths in the United States could be avoided through dietary modification, which includes an abundant intake of vegetables and fruits. Hence, consuming a diet rich in antioxidant vegetables will provide health-protective effects. Vegetables contain several phytochemicals which possess strong antioxidant activities. The antioxidant vegetables prevent from the cancer and other diseases by protecting cells from damage caused by 'free radicals'- highly reactive oxygen compounds.

In view of the above facts, some antioxidant vegetables (including spices and oils) commonly taken in the diet have been presented in this review article. Table 1 contains these vegetables in alphabetical order along with their Hindi/English names, parts used, main active components, antioxidant/anticancer and other biological activities.

Some Scientific Reports

A large body of scientific evidence indicates that an association exists between inadequate antioxidant status and increased risk for many diseases, including cancer, cardiovascular disease (CVD), diabetes, hypertension, leprosy, rheumatism, epilepsy, diseases of skin, eye, stomach, liver, lung, intestine, kidney and spleen, stroke, inflammation, fever, paralysis, asthma, diarrhoea, osteoporosis, osteoarthritis and peripheral arterial disease. Antioxidant vegetables have beneficial effect against several of these diseases. Vegetables contain compounds such as sulphoraphane that induces GSH transferase, there by helping detoxifying many types of carcinogens. Increased consumption of vegetables can increase the plasma antioxidant capacity and is associated with the lower risk of cancer and CVD (11). Vegetables are most effective against those cancers that involve epithelial cells such as cancers of lung, esophagus, stomach, colon, pancreas and cervix. The protective effect of vegetables has also been observed for hormone related cancers (8). The alcoholic extracts of dhaniya, haldi, karela and adrak were tested for tumour inhibitory effect. Haldi and karela were most effective against metastatic prostate cancer cell lines (7). Vegetables with the highest anticancer activity included lasun, soybean, pattagobhi, gajar; with a modest level of cancer-protective activity included piyaz, nibu, haldi, phoolgobhi, tamatar; with a low level of anticancer activity included khira (12).

The major groups of the vegetable phytochemicals which shows antioxidant activities include vitamins (A, C, E, K), carotenoids, terpenoids, flavonoids, polyphenols, saponins, enzymes and minerals. All of these prevent from the cancer and other diseases (3,5-6,9,13-14). Some phytochemicals which provide protection against cancer are: allyl sulphides present in lasun and piyaz; glucarates in nibu, began and alu; phytates, lignans, isoflavones and saponins in soybean; isothiocyanates in sarson, phoolgobhi and pattagobhi; and flavonoids, carotenoids and terpenoids in different vegetables. These phytochemicals block various hormone actions and metabolic pathways which are associated with the development of cancer (9).

Soybean is the contributing factor in the low incidence of breast, prostate, stomach, colon, rectum and lung cancers. Soybean seed contains isoflavonoids which inhibit the growth of hormone-dependent and hormone-independent cancer cells in culture. The isoflavones in soy inhibit the growth of human breast and prostate cancers (10). These isoflavones are the potent inhibitors of cholesterol synthesis, and the sterols (*e.g.*, β-sitosterol) and saponins in soy can block cholesterol absorption from the diet or increase cholesterol excretion from the body (9). Thus soybean may also protect from the heart disease. Nibu due to its flavonoid, tangeretin and nobiletin contents can potentially inhibits the tumor cell growth and can activate the detoxifying cytochrome P-450 enzyme system. Limonoids present in nibu inhibit tumor formation by stimulating glutathione S-transferase (GST) enzyme. Limonene (a terpenoid) content of nibu also possesses anticancer activity. Nibu has been used to inhibit the breast cancer cell proliferation, delay mammary tumorigenesis, and cure metastasis and leukemia cancers (9-10).

Table 1: Some antioxidant vegetables useful in cancer and other diseases.

Sl.No.	*Botanical (with Hindi or English) Name*	*Part Used*	*Main Active Components*	*Antioxidant/Anticancer and Other Biological Activities*	*References*
1.	*Abelmoschus esculentus* (L.) Moench (Bhindi, Lady's finger)	Fruit, seed	Carotene, vitamins (B, C), amino acids	Seeds inhibit cancer growth; diuretic, anti-catarrhal, stimulant, antispasmodic; given in dysentery.	17-20
2.	*Allium cepa* Linn. (Piyaz, onion)	Bulb	Diallyl disulphide, allicin, allin, quercetin antioxidant flavonoid, vitamins (C, E)	Diallyl disulphide inhibits stomach cancer; quercetin may cure lung and other cancers; analgesic, antidiabetic, antiinflammatory; given in jaundice, asthma, epilepsy, paralysis.	8-9,12,18, 20-21
3.	*Allium sativum* Linn. (Lasun, garlic)	Bulb	Sulphur compounds (diallyl sulphide, diallyl disulphide, allyl propyl disulphide), allicin	Allicin inhibits cancers of stomach, liver, colon, breast, endometrium; sulphur compounds inhibit cancer cells proliferation; antiinflammatory, antidiabetic; given in CVD.	9-10,12,18, 20
4.	*Beta vulgaris* Linn. (Chukandar, beet)	Whole plant	Vitamins (A, B, C), iodine, betanin	Given in liver disorders, paralysis, asthma, inflammation.	12,17,20
5.	*Brassica campestris* Linn. (Sarson, mustard)	Seed oil	Dithiolthiones, isothio-cyanates	Given in urinary problems, rheumatism, tumours/cancers	9,17-20
6.	*Brassica juncea* (*B. nigra*) Linn. (Rai, brown mustard)	Seed oil	Flavonol glycosides, dithiolthiones, isothiocyanate	Given in inflammation, rheumatism, paralysis.	9,12,20

Contd...

Table 1–*Contd...*

Sl.No.	*Botanical (with Hindi or English) Name*	*Part Used*	*Main Active Components*	*Antioxidant/Anticancer and Other Biological Activities*	*References*
7.	*Brassica oleracea* var. *botrytis* Linn. (Phoolgobhi, cauliflower)	Sprout (flowers)	Ascorbigen, vitamins (A, B, C), sulphoraphane, dithiolthiones, isothio-cyanate	Anticancer against cancers of bladder, lung, stomach, colon, rectum, breast; cardiotonic, antiinflammatory; given in biliousness, urinary diseases.	8-10,12, 18, 20
8.	*Brassica oleracea* var. *capitata* Linn. (Pattagobhi, cabbage)	Leaf	-do-	-do-	9-10,12,18, 20,23
9.	*Brassica rapa* Linn. (Shaljum, turnip)	Root, leaf, seed	Ascorbigen, vitamins (A, C), Ca	Diuretic; given in various tumours/ carcinomas.	12,18, 20-21
10.	*Capsicum annum* Linn. (Lal mirch, red chilli)	Fruit	Carotene, vitamin C, glutathione, flavonoids	Stimulant, stomachic; given in neuralgia, rheumatism, sore throat, diarrhoea	12,17-18, 20-21
11.	*Chenopodium album* Linn. (Bathua, Lamb quarters)	Whole plant, leaf, seed	Amino acids (leucine, isoleucine, lycine), vitamin C	Given in peptic ulcer, cardiac diseases.	12,17,20
12.	*Citrus limon* Linn. (Nibu, lemon)	Fruit	Vitamin C, flavonoid, flavone, limonoid, limonene (terpenoid), nobiletin, tangeretin, glucarate	Flavonoid, tangeretin, nobiletin, limonoid and limonene inhibit cancer growth; nibu inhibits human breast cancer; given in metastasis, leukaemia, rheumatism, diarrhoea, liver disorders.	9-10,17
13.	*Coriandrum sativum* Linn. (Dhaniya, coriander)	Leaf, fruit	Essential oil, vitamin C, carotene, borneao, limonene, α-pinene	Antioxidant, antiinflammatory, diuretic, stomachic, stimulant, hypolipidemic; given in rheumatism, jaundice, ulcer.	7,12,17, 19-20

Contd...

Table 1–*Contd...*

Sl.No.	*Botanical (with Hindi or English) Name*	*Part Used*	*Main Active Components*	*Antioxidant/Anticancer and Other Biological Activities*	*References*
14.	*Cucumis sativus* Linn. (Khira, cucumber)	Fruit, seed	Vitamin C, 14 α-methyl-phytosterol, amyrins (α, β)	Haemostatic, diuretic, tonic, astringent, antipyretic; given in bronchitis, jaundice, tumour, haemorrhage.	12,20
15.	*Curcuma longa* Linn. (Haldi, turmeric)	Rhizome (Tuber)	Curcumin, β pinene, camphene, eugenol, curcuminoids, β sitosterol	Active against esophagus, colon, liver, bladder and prostate cancers, leukaemia, fibrosarcoma, stomach papilloma; hepatoprotective, hypo-lipidemic cardiotonic, antidiabetic.	7,9,12, 18-21,25
16.	*Daucus carota* Linn. (Gajar, carrot)	Root, leaf	Carotene, flavonoid, carotenoid, glycoside	Given in leprosy, chest troubles, tumours, bronchitis, jaundice	12,19-20
17.	*Dolichos lablab* Linn. (Sem, butter bean)	Pod, seed	Proteins, amino acids, vitamins (A, B, C)	Stomachic, antiseptic; given in abdominal pain, diarrhoea, vomiting	12, 20
18.	*Glycine javanica* Linn. (Soybean)	Seed	Phytates, saponins, phytosterols, lignans, isoflavonoids, isoflavones	Lowers the risk of breast, prostate, colon, stomach, rectum and lung cancers; isoflavonoids and isoflavones inhibit growth of many cancers.	9-10
19.	*Lycopersicon esculentum* Mill. (Tamatar, tomato)	Fruit	Vitamins (A, B, C), essential amino acids, lycopene	Antioxidant, anticancer, intestinal antiseptic, cholagogue, stimulant; given in lung and heart diseases.	10,17-20
20.	*Mentha spicata* Linn. (Pudina, garden mint)	Leaf	Limonene, flavonoid, sesquiterpene	Antioxidant, antiinflammatory, antiulcero-genic; given in asthma.	12,17, 19-20
21.	*Momordica charantia* Linn. (Karela, bitter gourd)	Leaf, fruit, seed	Linoleic and palmitic acids, momordin, vitamins	Active against colon, breast, bladder and prostate cancers, lymphoma, leukaemia; antidiabetic; given in asthma, bronchitis	7,12,18, 20, 25

Contd...

Table 1–*Contd...*

Sl.No.	*Botanical (with Hindi or English) Name*	*Part Used*	*Main Active Components*	*Antioxidant/Anticancer and Other Biological Activities*	*References*
22.	*Murraya koenigii* (L.) Spreng (Mitha neem, curry-leaf tree)	Leaf	Carotene, vitamins, alkaloids	Stomachic, carminative; given in hysteria, diarrhoea, dysentery.	12,17-18, 20
23.	*Nelumbo nucifera* Gaertn. (Kamal, lotus)	Whole plant	Alkaloids, vitamins, quercetin flavonoid	Antioxidant, antipyretic, demulcent, diuretic; given in neurasthenia, uterine hemorrhage, leprosy, liver diseases.	12,17,20
24.	*Phaseolus vulgaris* (French bean)	Fruit	Amino acids, anthocyanin, quercetin	Antioxidant, anticancer; urine flow stimulator; given in diabetes.	9,12,20
25.	*Raphanus sativus* Linn. (Muli, radish)	Root, leaf	Vitamin C, anthocyanin, glucosinolates	Cardiotonic, analgesic, laxative; given in asthma, dysuria, calculi, bronchitis, problems of liver and heart	12,17, 20-21
26.	*Solanum melongena* Linn. (Began, brinjal)	Fruit, seed	Solasodine, glucarate	Stimulant, antiglosstic; given in cough, indigestion, liver complaints	9,12,17,20
27.	*Solanum tuberosum* Linn. (Alu, potato)	Tuber	Solanine, solasodine, lutein, vitamins (A, C), glucarate	Diuretic, aperient, nutritive; given in leg ulcer, cough, weakness.	9,12,17, 20,24
28.	*Spinacia oleracea* Linn. (Palak, spinach)	Whole plant	Vitamins, carotene, lecithin, amino acids	Antibacterial, laxative; given in inflammation of lung, liver and bowel	12,17-18
29.	*Trigonella foenum-graecum* Linn. (Methi, fenugreek)	Leaf, seed	Choline, trigonelline, saponin, amino acids, vitamins, quercetin	Given in various cancers, bronchitis, inflammatory affections, rheumatism.	7,12, 17-18, 20
30.	*Zingiber officinale* Linn. (Adrak, ginger)	Rhizome	Camphene, gingerol, zingiberene, borneol, cineol, curcumins, proteins	Antioxidant, anticancer, antipyretic, analgesic, antiinflammatory, astringent; given in swelling, tumours, colic, diabetes, heart disease.	7, 9,12, 18, 20, 25-26

Vegetables also contain a variety of isoprenoid compounds that exhibit antitumour/anticancer activities. These compounds include tocotrienols (related to tocopherols) and terpenoids (*e.g.*, limonene, geranoil, menthol and carvone). Overall, these compounds increase tumour latency and decrease tumour multiplicity. These compounds also elicit a significant reduction in total and LDL cholesterol levels, there by reducing the risk of heart disease (15). There are many vegetable pigments such as flavonoids, carotenoids and anthocyanins. These pigments protect us from various diseases. Flavonoids extend the activity of vitamin C; act as antioxidant; protect LDL cholesterol from oxidation to the unsafe cholesterol oxides; inhibit platelet aggregation; and have antiinflammatory and antitumour activities (16). Quercetin (a flavonoid) chiefly present in red and yellow piyaz and French bean possesses both anticarcinogenic activity and the ability to inhibit LDL oxidation. Anthocyanins inhibit cholesterol synthesis and thus provide protection against heart disease. The carotenoids are powerful antioxidants that provide protection against oxidative damage, and stimulate immune function. Persons with high levels of serum carotenoids have a reduced risk of heart disease and cancer. In addition, a variety of phenolic compounds (*e.g.*, caffeic, ellagic and ferulic acids, sesamol, and vanillin) are present in vegetables. These exhibit antioxidant and anticancer activities and inhibit atherosclerosis. Adrak contains phenolic compounds (gingerol) which have antioxidant activity that is even greater than α-tocopherol (vitamin E). Compounds that stimulate the activity of GST are considered as inhibitors of cancer. Substances that stimulate GST activity are phthalides in celery seeds; sulphides in piyaz and lasun; dithiolthiones and isothiocyanates in sarson, rai, phoolgobhi and pattagobhi; limonoids in nibu; and curcumins in adrak and haldi (9).

Conclusion

Vegetables are the most important part of our food which are consumed daily but unfortunately very few of us know that they provide significant protection against many cancers and other diseases. Consumption of vegetables is widely accepted as lowering the risk of several cancers. The antioxidant vegetables prevent from cancer and other diseases by protecting cells from damage. Hence, consuming a diet rich in antioxidant vegetables will certainly provide health-protective effects. People wishing to reduce the risk of cancer must eat several portions of vegetables and several pieces of fruits every day. Though many scientific studies have been done to explore out the antioxidant, anticancer and other biological activities of several vegetables; however, there is still a great need and scope for the researchers, biologists, pharmacologists and pharmaceutical scientists to derive novel anticancer or other drugs from such vegetables.

References

1. Madhuri, S. and Pandey, Govind (2008). Some dietary agricultural plants with anticancer properties. Pl. Arch., 8(1): 13-16.
2. Pandey, Govind and Madhuri, S. (2006). Medicinal plants: better remedy for neoplasm. Indian Drug, 43(11): 869-874.
3. Pandey, Govind and Madhuri, S. (2008). Some anticancer agents from plant origin. Pl. Arch., 8(2): 527-532.
4. Sivalokanathan, S., Ilayaraja, M. and Balasubramanium, M.P. (2005). Efficacy of *Terminalia arjuna* (Roxb.) on N-nitrosodiethylamine induced hepatocellular carcinoma in rats. Indian J. Exp. Biol., 43: 264-267.
5. Kathiresan, K., Boopathy, N.S. and Kavitha, S. (2006). Costal vegetation- An underexplored source of anticancer drugs, Nat. Prod. Rad., 52: 115-119.
6. Madhuri, S. and Pandey, Govind (2009). Some anticancer medicinal plants of foreign origin. Current Science, 96(6): 779-783.
7. Rao, K.V.K., Schwartz, S.A., Nair, H.K., Aalinkeel, R., Mahajan, S., Chawda, R. and Nair, M.P.N. (2004). Plant derived product as a source of cellular growth inhibitory phytochemical on PC-3M, DU-145 and LNCaP prostate cancer cell lines. Current Science, 87(11): 1585-1588.
8. Vecchia, C.L. and Tawani, A. (1998). Fruits, vegetables and human cancer. Eur. J. Cancer, 7: 3-8.
9. Craig, W.J. (2006). Phytochemicals: Guardians of our health website.
10. Heber, D. (2004). Vegetables, fruits and phytoestrogens in the prevention of diseases. J. Postgrad. Med., 50: 145-149.
11. Polidori, M.C. (2003). Antioxidant micronutrients in the prevention of age-related diseases. J. Postgrad. Med., 49: 229-235.

12. Kaur, C. and Kapoor, H.C. (2002). Antioxidants activity and total phenolic content of some Asian vegetables. International J. Food Sci. Tech., 37: 153-161.

13. Madhuri, S. and Pandey, Govind (2010). Effect of ProImmu, a herbal drug on estrogen caused uterine and ovarian cytotoxicity. biomed, 5(1): 57-62.

14. Ray, G. and Hussan, S.A. (2002). Oxidant, antioxidant and carcinogenesis. Indian J. Exp. Biol., 40: 1213-1232.

15. Elson, C.E. and Yu, S.G. (1994). The chemoprevention of cancer by mevalonate-derived constituents of fruits and vegetables. J. Nutr., 124: 607-614.

16. Kanner, J., Frankel, E. and Granit, E. (1994). Natural antioxidants in grapes and wines. J. Agric. Food Chem., 42: 64-69.

17. Kaushik, P. and Dhiman, A.K. (1999). *Medicinal Plants and Raw Drugs of India.* Bishen Singh Mahendra Pal Singh, Dehradun, India.

18. CSIR (1986). *The Useful Plants of India.* Council of Scientific and Industrial Research, New Delhi, India.

19. IDMA (2002). *Indian Herbal Pharmacopoeia.* Mumbai, India.

20. Prajapati, N.D., Purohit, S.S., Sharma, A.K. and Kumar, T. (2003). *A Hand Book of Medicinal Plants.* Agrobios, India.

21. Jadeja, B.A., Odedra, N.K., Solanki, K.M. and Baraiya, N.M. (2006). Indigenous animal healthcare practices in district Porbandar, Gujarat. Indian J. Trad. Knowledge, 5(2): 253-258.

22. Arivazhagan, S., Velmurugan, B., Bhuvaneswari, V. and Nagini, S. (2004). Effect of aqueous extract of garlic (*Allium sativum*) and neem (*Azadirachta indica*) leaf containing blood oxidant-antioxidant status during experimental gastric carcinogesis. J. Med. Food, 7(3): 334-339.

23. Beecher, C.W.W. (1994). Cancer preventive properties of varieties of *Brassica oleracea*: A review. *Am. J. Clin. Nut.*, 59 (Suppl.): 1166-1170.

24. AI-Saikhon, M.S., Howard, L.R. and Miller, J.C. (1995). Antioxidant activity and total phenolics in different genotype of potato (*Solanum tuberosum*). J. Food Sci., 60: 341-343.

25. Aruna, K. and Sivaramkrishanan, V.M. (1990). Plant products as protective agents against cancer. Indian J. Exp. Biol., 28: 1008-1011.

26. Gazzani, G., Papetti, A., Massolini, G. and Daglia, M. (1998). Antioxidant and pro-oxidant activity of water soluble component of some common diet vegetable and the effect of thermal treatment. J. Food Chem., 6: 4118- 4122.

Some Important Anticancer Herbs: A Review

Govind Pandey*

Ex-Professor/Principal Scientist and Head of Pharmacology (Pharmacy); presently Officer-In-Charge of Rinder Pest (Animal Husbandry Department, Govt. of MP), Jabalpur Division, Jabalpur, MP, India

**Officer-In-Charge of Rinder Pest (Animal Husbandry Department, Govt. of MP), Jabalpur Division, Jabalpur-482001, MP, India;*

E-mail: drgovindpandey@rediffmail.com

ABSTRACT

A great deal of pharmaceutical research has considerably improved the quality of herbal drugs used against various types of cancer. With the advanced knowledge of molecular science and the refinement in isolation and structure elucidation techniques, we are in a much better position now to identify various anticancer herbs. Scientists all over the world are concentrating on the use of herbs to boost immune system of the body against cancer. Scientists have contributed for a number of years to identify hundreds of anticancer herbs, and developed various herbal formulations from their active principles that inhibit growth and spread of cancer without any side effect. Such herbs possess anticancer, immuno-enhancing, antiangiogenesis, antioxidant and antimutagenic properties. They inhibit growth

Article Received on: 12/05/11 Revised on: 20/06/11 Approved for publication: 08/07/11.

[Paper published in: International Research Journal of Pharmacy, 2011; 2(7): 45-52].

and spread of cancer by modulating the activity of hormones, enzymes and other biological factors. The therapeutic effect of these herbs is executed by the complex synergistic interaction among their various active principles. Some important anticancer herbs have been discussed here.

Keywords: *Anticancer activities, Cancers, Herbs/medicinal plants, Phytoconstituents.*

Introduction

A large number of medicinal plants act as anticancer herbs in experimental and/or clinical cancers/tumours of various organs. Some of those cancers are sarcoma, leukaemia, lymphoma and carcinoma[1–2]. This review article contains 35 anticancer herbs (anticancer medicinal plants) which have been described ahead. The general data of these plants have been collected from some books authored by different authors[3–6]. However, the particular phytoconstituents (phytochemicals) present in these plants, their mechanism of action and uses against the various cancers as reported by different authors have been cited under each medicinal plant/herb.

Medicinal plants are known to have versatile immunomodulatory and antioxidant properties, leading to anticancer activities. These act by stimulating both non-specific and specific immunity. They may promote host resistance against infection by re-stabilizing body equilibrium and conditioning the body tissues[1–2,7–10]. Many reports describe that the anticancer activity of the medicinal plants is because of the presence of certain phytoconstituents, which possess strong antioxidant activities. The antioxidants may prevent and cure the cancer and other diseases by protecting cells from damage caused by 'free radicals' the highly reactive oxygen compounds. Thus, consuming a diet rich in antioxidant plant foods will provide a milieu of phytoconstituents, non-nutritive substances in plants that possess health-protective effects. The main phytoconstituent antioxidants with anticancer activity include vitamins (*e.g.*, A, C, E and K), carotenoids or carotene, terpenoids, flavonoids, polyphenols (*e.g.*, ellagic acid, gallic acid and tannins), flavonoids (*e.g.*, quercetin, anthocyanins, catechins, isocatechins, flavones, flavonones and isoflavones), enzymes (*e.g.*, superoxide dismutase, catalase and glutathion peroxidase), minerals (*e.g.*, selenium, copper, manganese, zinc, chromium and iodine), polysaccharides, alkaloids, saponins, lignins, xanthones and certain pigments[1–2,7–12].

Some Important Medicinal Plants or Phytoconstituents with Anticancer Activities

Aegle marmelos Correa ex Roxb. (Bel; Family: Rutaceae)

Lupeol, isolated from pulp and seeds of *A. marmelos*, possesses strong anticancer activity against breast cancer, malignant lymphoma, malignant melanoma, malignant ascites and leukaemia. *A. marmelos* possesses significant antioxidant activity and reduces side effects of chemotherapy and radiotherapy.

Allium cepa Linn. (Piyaz/Onion; Family: Liliaceae/Alliaceae)

Diallyl disulphide, quercetin flavonoid, allicin, allin and vitamins (C, E), isolated from bulb of *A. cepa,* detoxify carcinogen, inhibit *Helicobacter pylori* and arrest cell cycle from S to G2M phase. Diallyl disulphide inhibits stomach cancer, and quercetin may cure lung and other cancers[8].

Allium sativum Linn. (Lasun/garlic; Family: Liliaceae/Alliaceae)

Sulphur compounds (diallyl sulphide, diallyl disulphide, allyl propyl disulphide) and allicin have been isolated from *A. sativum* bulb. Allicin inhibits growth of stomach, liver, colon, breast and endometrium cancers; while sulphur compounds inhibit the cancer cells[8].

Aloe vera Tourn. ex Linn./*A. barbadensis* Mill. (Ghee-Kunwar/Indian Aloe; Family: Liliaceae)

Acemannan (a polysaccharide) isolated from root, pulp, leaves or aerial parts of *A. vera* stimulates immune system and possesses significant anticancer activity. Emodin and lectins isolated from this herb exhibit strong anticancer and immunoenhancing activities. Aloe-emodin inhibits growth and spread of stomach cancer and various sarcomas by inducing apoptosis. Aloe-emodin has selective anticancer activity against neuroectodermal tumours. Alexin B isolated from *A. vera* possesses strong anticancer activity against leukaemia. Its polysaccharides have strong immunoenhancing and anticancer properties. *A. vera* contains "super carbohydrates" that protect against many cancers, particularly the liver cancer. This herb prevents genesis of cancer, regresses growth of cancer and prevents metastasis of cancer. *A. vera* stimulates immune system response of the body by activating macrophages and releasing cytokines such as interferon, interleukin and tumour necrosis factor. *A. vera* has an extraordinary antioxidant profile and reduces side effects of chemotherapy and radiotherapy. Its leaves contain glycosides-anthracene derivatives or hydroxyanthraquinone derivatives[7].

Alpinia galanga **Willd. (Barakulanjan; Family: Zingiberaceae)**

Acetoxy-chavicol-acetate, isolated from *A. galanga*, possesses significant anticancer activity against cancers of breast, lung, stomach, colon and prostate, multiple myeloma, and leukaemia. Pinocembrin isolated from this herb inhibits growth and spread of colon cancer by arresting cell proliferation and inducing apoptosis. Galangin, a flavonoid isolated from *A. galanga*, possesses strong anticancer, antioxidant, antimutagenic and antiinflammatory properties. Galangin protects against breast and prostate cancers.

Andrographis paniculata **Wall. ex Nees (Kiryat/Kalmegh/Creat; Family: Acanthaceae)**

Andrographolide (active diterpine component), isolated from whole plant of *A. paniculata*, has immunoenhancing and strong anticancer activity against cancers of breast, ovary, stomach, colon, prostate, kidney and nasopharynx, malignant melanoma, and leukaemia. Andrographolide has been observed to be potential enhancers of immune system functions such as production of white blood cells (the defense cells of our body), release of interferon (an antiviral factor) and activity of the lymphatic system (the seat of defense system). Andrographolide exerts direct anticancer activity on cancer cells by arresting G0/G1 phase of cell-cycle and inducing apoptosis. Dichloromethane fraction of methanolic extract of *A. paniculata* has strong anticancer activity against colon cancer. *A. paniculata* extract is cytotoxic (cell-killing) against cancer cell as seen in human epidermoid carcinoma of skin, lining of nasopharynx and lymphocytic leukaemia cells. The chemoprotective potential of *A. paniculata* against chemotoxicity, including carcinogenicity was observed in mice. Thus, *A. paniculata* possesses anticancer, immunostimulant, antioxidant, anti-HIV, antiinflammatory and antihepatotoxic properties. It enhances the activity of protective liver enzymes and reduces side effects of chemotherapy and radiotherapy. *A. paniculata* also contains flavonoid and andrographin[9].

***Aphanamixis polystachya* (Wall.) Parker/*Amoora rohituka* Wight and Arn. (Harinhara/Amoora; Family: Meliaceae)**

Amooranin (a triterpene acid), isolated from *A. polystachya* stem bark, inhibits growth and spread of breast and cervical cancers by arresting G2/M phase of the cell cycle and by inducing apoptosis. Amooranin and its derivatives are effective in both chemotherapy-sensitive and chemotherapy-resistant cancers. Amooranin has the ability to overcome (reverse) multidrug- resistance in breast cancer, colon cancer and leukaemia.

***Azadirachta indica* A. Juss./*Melia azadirachta* Linn. (Neem; Family: Meliaceae)**

Stem bark, leaf and flower of *A. indica* contains about 40 different active principles, known as limonoids, which exhibit immunoenhancing, antioxidant, antimutagenic, anticancer and antimetastatic, antiinflammatory, hepatoprotective, antiulcer, antifungal and antiviral activities. Limonoids regress growth and spread of various cancers, *e.g.*, cancers of breast, lung, liver, stomach, prostate and skin. Nimbolide, a natural triterpenoid, isolated from *A. indica* leaves and flowers inhibits growth and spread of various cancers, including colon cancer, malignant lymphoma, malignant melanoma and leukaemia by inducing apoptosis (programmed cell death), a process that directs the body's immune cells to identify and destroy cancer cells. Nimbolide also prevents metastasis of cancer. Ethanolic extract of *A. indica* inhibits growth and spread of prostate cancer by inducing apoptosis and its antiandrogenic effect. This herb reduces side effects of chemotherapy and radiotherapy. *A. indica* also contains polyphenolic myoinositol, dexamethasone, tannin, β-sitosterol, nimbin, quercetin and carotene[2,9].

***Bauhinia variegata* Linn. (Kachnar; Family: Caesalpiniaceae)**

Cyanidin glucoside, malvidin glucoside, peonidin glucoside and kaempferol galactoside, isolated from root, stem bark and flower of *B. variegata,* inhibit growth and spread of various cancers, *e.g.*, cancers of breast, lung, liver, oral cavity and larynx, and malignant ascites. *B. variegata* also possesses significant hepatoprotective activity.

***Berberis vulgaris* Linn. (Kashmal; Family: Berberidaceae)**

B. vulgaris root bark contains berberine, berbamine, chelidonic acid, citric acid, columbamine, hydrastine, isotetrandrine, jacaranone, magnoflorine, oxycanthine and palmatine. Berberine (an isoquinoline alkaloid) possesses anticancer, immunoenhancing, antioxidant and antiinflammatory properties. Berberine arrests cancer cell cycle in G1-phase and induces apoptosis, and hence it possesses strong anticancer activity against prostate cancer, liver cancer and leukaemia. Berberine interferes with P-glycoprotein in chemotherapy-resistant cancers. It also increases the penetration of some chemotherapy drugs through the blood-brain barrier, thereby enhancing their effect on intracranial tumours. *B. vulgaris* root bark contains three phenolic compounds, viz., tyramine, cannabisin-G and lyoniresinol. Cannabisin-G and lyoniresinol exhibit strong antioxidant activity. Cannabisin-G protects against breast cancer. *B. vulgaris* also inhibits growth of stomach and oral cavity cancers.

***Catharanthus roseus* G. Don/*Vinca rosea* Linn./*Lochnera rosea* (Linn.) Reichb. (Sadabahar/Madagascar Periwinkle; Family: Apocynaceae)**

C. roseus whole plant contains more than 70 alkaloids, known as vinca alkaloids such as vinblastine, vincristine and their derivatives. Vinca alkaloids arrest cancer cell proliferation by binding to tubulin in the mitotic spindle, *i.e.*, they inhibit microtubule formation and arrests mitosis in metaphase Vinca alkaloids also induce apoptosis (programmed cell death) and inhibit angiogenesis (formation of new blood vessels). These alkaloids inhibit growth and spread of various cancers, including breast, ovary, cervix, lung, colon, rectum, kidney and testis cancers, neuroblastoma, Hodgkin's disease, malignant lymphoma, multiple myeloma, various sarcomas, rhabdomyosarcoma, and leukaemia[2,7].

***Curcuma longa* Linn./*C. domestica* Valeton (Haldi/Turmeric; Family: Zingiberaceae)**

Curcumin (Diferuloyl methane) and curcuminoids, isolated from *C. longa* rhizome (tuber) suppress cancer at every step, *i.e.*, initiation, growth and metastasis. Curcumin (pigment colour of haldi) arrests the cancer cells proliferation in G2/S phase and induces apoptosis (programmed cell death). Curcumin has shown antiinflammatory, antitumour and antioxidant properties. It inhibits angiogenesis, a crucial step in the growth and metastasis of cancer. Curcumin and genistein (isolated from *Glycine max*) act synergistically to inhibit growth and spread of oestrogen-positive breast cancer. Curcumin acts even in multidrug-resistant breast cancers. Curcumin suppresses adhesion of cancer cells, thus preventing metastasis.

It inhibits growth and spread of various cancers, including that of breast, lung, oesophagus, liver, colon, prostate, head, neck and skin. Curcumin is particularly effective in radiotherapy-resistant prostate cancer. It is effective even in advanced stages of cancer. Curcumin showed chemopreventive effect against N-nitrosodiethylamine/phenobarbital induced-hepatocarcinogenesis in wistar strain male albino rats. It also protects from stomach and colon cancers. *C. longa* rhizome is also antimutagenic, antioxidant, immunostimulant, antiinflammatory, radioprotective, stimulant, carminative, alterative, blood purifier, hepatoprotective, antiperiodic and tonic. Rhizomes are also effective in colon, bladder and prostate cancers, intravesical tumour, fibrosarcoma, hepatocellular carcinoma, oesophagal carcinogenesis, leukaemia, stomach papilloma and solid tumours[2,13].

***Emblica officinalis* Gaertn./*Phyllanthus emblica* Linn. (Amla/Amlika/ Indian Gooseberry; Family: Euphorbiaceae)**

E. officinalis fruit contains ellagic acid, gallic acid, quercetin, kaempferol, emblicanin, flavonoids, glycosides and proanthocyanidins. It is valued for its unique tannins and flavanoids, which possess powerful immunomodulatory, antioxidant and anticancer activities. Ellagic acid is a powerful antioxidant and has the ability to inhibit mutations in genes. Ellagic acid also repairs chromosomal abnormalities. Quercetin has hepatoprotective effect. Emblicanins A and B (tannins) possess strong antioxidant and anticancer properties. *E. officinalis* inhibits growth and spread of various cancers, including that of the breast, uterus, pancreas, stomach and liver, and malignant ascites. *E. officinalis* is an excellent rejuvenator and antioxidant herb. It is highly nutritious and an important source of vitamin C (a powerful antioxidant), phyllembic acid, lipid, emblicol, colloidal complexes, micic acid, amino acids and minerals. *E. officinalis* protects against many cancers, particularly the liver cancer. It reduces side effects of chemotherapy and radiotherapy. Amla fruits are also acrid, cooling, aperient, refrigerant, astringent, diuretic and laxative. The dried fruits are useful in inflammation, haemorrhage, cough, diarrhoea, anaemia, jaundice and dyspepsia. Flowers of *E. officinalis* are cooling, refrigerant and aperient, while root and bark are astringent. *E. officinalis* seeds are used for asthma, bronchitis and biliousness. Due to rich vitamin C (ascorbic acid or ascorbate), amla is successfully used in the treatment of human scurvy. Phyllembin, from fruit pulp identified as ethyl gallate; tannin from fruit, bark and leaves; fixed oil, essential oil and phophatides from seeds; and leucodelphinidin from bark of *E. officinalis* have also been isolated. Amla fruit contains 18 compounds that inhibit growth of gastric, uterine and breast cancers. It enhances natural killer cell activity in various tumours. Its extract reduced the ascites and solid tumours induced by Dalton's lymphona ascites cells in mice. The extract also increased the life span of tumour bearing animals[2,14].

***Fragaria vesca* Linn. (Strawberry; Family: Rosaceae)**

Flavonoid, tannin, borneol, ellagic acid, vitamin C, bioflavonoid and chalcone of *F. vesca* leaf and fruit possess antioxidant and anticancer activities. *F. vesca* acts against various cancers[1–2].

***Ginkgo biloba* Linn. (Maidenhair Tree; Family: Ginkgoaceae)**

Ginkgetin and ginkgolides (A and B), isolated from *G. biloba,* inhibits growth and spread of various aggressive cancers such as invasive oestrogen-receptor negative breast cancer, glioblastoma, hepatocellular carcinoma, and cancers of ovary, colon, prostate and liver by inducing apoptosis. *G. biloba* extract is well known for its antioxidant activity. It also reduces side effects of chemotherapy and radiotherapy.

***Glycine max* Merrill/*G. soja* Sieb. and Zucc./*G. hispida* Maxim/*Soja max* Piper (Bhat/Soybean/Soyabean/Soya; Family: Papillionaceae/Fabaceae)**

Isoflavones (*e.g.*, genistein and daidzein) and saponins, isolated from *G. max* beans (seeds), inhibit growth and spread of various cancers such as cancers of breast, uterus, cervix, ovary, lung, stomach, colon, pancreas, liver, kidney, urinary bladder, prostate, testis, oral cavity, larynx, and thyroid. *G. max* is also effective in nasopharyngeal carcinoma, skin cancer, malignant lymphoma, rhabdomyosarcoma, neuroblastoma, malignant brain tumours and leukaemia. Isoflavones and saponins possess wide ranging anticancer properties like inhibition of cancer cell proliferation, promotion of cell differentiation and induction of apoptosis. Genistein acts by blocking angiogenesis, acts as a tyrosine kinase inhibitor (the mechanism of action of many new cancer drugs) and induces apoptosis. Genistein is an excellent intracellular antioxidant, which blocks the supply of oxygen and nutrients to cancer cells, thus killing them by starving. Genistein and quercetin have synergistic anticancer effect against ovarian carcinoma. Saponins decrease invasiveness of glioblastoma cells. Anthocyanins of *G. max* induce apoptosis in leukemic cells. *G. max* protects against many cancers, including that of colon, lung and ovary.

***Glycyrrhiza glabra* Linn. (Mulathi/Mulhatti/Licorice/Liquorice); Family: Papilionaceae/Fabaceae/Leguminosae)**

Flavonoids (*e.g.*, flavones, flavonols, isoflavones, chalcones, licochalcones and bihydrochalcones), derived from root, rhizome or whole plant of *G. glabra* possess strong anticancer, antioxidant, antimutagenic, antiulcer, anti-HIV and hepatoprotective properties. Licochalcone-A inhibits growth and spread of various cancers particularly the androgen-refractory prostate cancer by inducing apoptosis and arresting cancer cells division. Licoagrochalcone possesses strong anticancer activity against cancers of breast, lung, stomach, colon, liver and kidney, and leukaemia. Triterpenoid saponins (*e.g.*, glycyrrhizin and glabranin) isolated from *G. glabra* inhibits growth and spread of lung cancer and fibrosarcomas. Glycyrrhizic acid isolated from this herb protects against aflatoxins (powerful fungal carcinogens of the liver). *G. glabra* also contains coumarin, triterpene sterol (β amerin stigmasterol), eugenol, indole, glycyrrhetinic acid, chalcone glycosides (viz., isoliquiritin and neoisoliquiritin), and liquiritoside (a flavonoside). This herb stimulates immune system response of the body and protects against colon cancer and oestrogen-positive breast cancer. The rhizomes and roots are also tonic, expectorant, demulcent, laxative and emollient, and used in genito-urinary diseases, coughs, sore throat, catarrhal affections and in scorpion-sting. Licorice is an extract prepared from the dried roots and stems of *G. glabra*. For more than three thousand years, licorice has been used to treat cancer, hepatitis and some other diseases. Antitumour and antimetastatic effects of cyclophosphamide are potentiated by licorice extract[7,9].

***Malus domestica* Borkh./*M. pumila* Mill./*M. communis* DC/*M. sylvestris* Hort., non Mill./*Pyrus malus* Linn. in part (Seb/Sev/Apple; Family: Rosaceae)**

Fruit (Apple) of *M. domestica* possesses antioxidant and anticancer activities, and may be useful in various cancers[2].

***Morinda citrifolia* Linn. (Al/Ach/Bartundi/Noni; Family: Rubiaceae)**

M. citrifolia has 23 different phytochemicals, including five vitamins and three minerals. *M. citrifolia* heartwood contains active constituents as anthraquinones, *viz.*, damnacanthol, rubiadin-methyl ether, alizarin, morindone and anthragallol-2, 3-dimethyl ether. Damnacanthol, NB10 and NB11, isolated from *M. citrifolia* fruit, possess strong anticancer activity against various cancers, particularly lung cancer and sarcomas. *M. citrifolia* possesses strong antioxidant, hepatoprotective and immunoenhancing properties. The flowers on ethanolic extraction yielded acacetin 7-O-(-D(+)- glucopyranoside; 5, 7-dimethyl-apigenin-4′-O-(-D(+) galactopyranoside and a new anthraquinone glycoside. The fruit juice of *M. citrifolia* showed antitumour activity against intraperitoneally implanted Lewis lung carcinoma in syngenic mice. Noni fruit extract acts indirectly on cancer cells by enhancing host immune system. There is a polysaccharide compound (6-D-glucopyranone pentaacetate) found in the Noni which increases the ability of immune system to produce chemicals that enhance the killing power of the white blood cells against cancer. Noni fruit provides a safe and effective way to increase xeronine levels, which exert a crucial influence on cell health and body protection. Its fruit contains proxeronine (a precursor of xeronine), which initiates the release of xeronine in the intestinal tract after it comes in contact with a specific enzyme (present in the fruit). Thus, xeronine is an alkaoid to which the body produces in order to activate enzymes, so they can function properly. This particular alkaloid has never been found because the body makes it, immediately uses it, and then breaks it down. Xeronine is so basis to the functioning of proteins, we would die without it. Its absence can cause many kinds of illness. Noni, which is probably the best source of proxeronine, acts as an immuno-stimulant, inhibits the growth of certain tumours, enhances and normalizes cellular functions, and boosts tissue regeneration[15].

Nigella sativa Linn. (Kalonji/Kalajira/Black Cumin; Family: Ranunculaceae)

Thymoquinone and dithymoquinone, isolated from *Nigella sativa* seeds, have strong anticancer activity against various cancers, including cancers of colon, prostate, pancreas and uterus, malignant ascites, malignant lymphoma, malignant melanoma, sarcomas, and leukaemia. Thymoquinone is effective in both hormone-sensitive and hormone-refractory prostate cancers. *N. sativa* kills cancer cells by binding to the asialofeutin (lectin) on the surface of cancerous cells, causing their aggregation and clumping. It also possesses immunoenhancing and anti-inflammatory properties. It protects against liver cancer. *N. sativa* enhances immune function of the body and reduces side effects of chemotherapy and radiotherapy.

***Ocimum sanctum* Linn. (Tulsi/Sacred Basil/Holy Basil; Family: Labiatae/ Lamiaceae)**

O. sanctum leaves contain volatile oils (comprising of eugenol and methyl eugenol), linoleic acid, oleanic acid, rosmarinic acid, and flavonoids or phenolic compounds as antioxidants (*e.g.*, orientin, vicenin, cirsilineol, cirsimaritin, isothymusin, isothymonin and apigenin). The volatile oils also contain carvacrol and sesquiterpine hydrocarbon caryophyllene. Ursolic acid, apigenin, luteolin, apigenin-7-O-glucuronide, luteolin-7-O-glucuronide, orientin and molludistin have also been isolated from the leaf. It also contains a number of sesquiterpenes and monoterpenes, *viz.*, bornyl acetate, β-elemene, neral, α- and β-pinenes, camphene, campesterol, cholesterol, stigmasterol, and β-sitosterol. Eugenol, orientin and vicenin inhibit growth and spread of various cancers such as breast cancer, liver cancer and sarcomas, particularly fibrosarcoma by blocking supply of oxygen and nutrients to cancer cells and killing them by starving. Ursolic acid has immunoenhancing and tissue-protective properties. Polysaccharides isolated from it have antioxidant and radioprotective properties. *O. sanctum* possesses antioxidant, antitumour and immunomodulatory activities and protects against various cancers, particularly breast cancer and reduces side effects of chemotherapy and radiotherapy. The alcoholic extract (AlE) of *O. sanctum* leaves has a modulatory influence on carcinogen metabolizing enzymes such as cytochrome P 450, cytochrome b_5, aryl hydrocarbon hydroxylase and glutathione S-transferase (GST), which are important in detoxification of carcinogens and mutagens. *O. sanctum* significantly decreased the incidence of benzo(a)pyrine induced neoplasia of fore-stomach of mice and 3′-methyl-4-dimethylaminoazo-benzene induced hepatomas in rats. The AlE of the leaves of OS was shown to have an inhibitory effect on chemically induced skin papillomas in mice. Leaf extract of *O. sanctum* blocks or suppresses the events associated with chemical carcinogenesis by inhibiting metabolic activation of the carcinogen. The anticancer activity of *O. sanctum* was observed in Swiss albino mice bearing Ehrlich ascites carcinoma and sarcoma 180 tumours. Besides above, *O. sanctum* herb is also useful as antibacterial, antiviral, antifungal, antiprotozoal, antimalarial, anthelmentic, antidiarrhoeal, analgesic, antipyretic, antiinflammatory, antiallergic, antihypertensive, cardioprotective, central nervous system depressant, memory enhancer, antihypercholesterolaemic, antihepatotoxic, antidiabetic, antiasthmatic, antithyroidic, chemopreventive, radioprotective, antifertility, antiulcer, antiarthritic, adaptogenic/antistress, anticataract, antileucodermal, anticoagulant, aromatic, stomachic, carminative, demulcent, diaphoretic, diuretic, expectorant, alexiteric, vermifuge and febrifuge. Tulsi is a popular home remedy for many illnesses and everyday ailments like common cold, headache, cough, flu, earache, fever, colic pain, sore throat, bronchitis, flatulence, migraine headaches, insomnia, arthritis, night blindness, wound, liver diseases, catarrhal fever, otalgia, lumbago, hiccough, ophthalmia, digestive disorders, genitourinary disorders, diarrhea, influenza, skin diseases, various forms of poisoning and psychosomatic stress disorders[2,7,16].

Oldenlandia diffusa Roxb./_Hedyotis diffusa_ Willd. (Family: Rubiaceae)

O. diffusa is a native of China. The whole plant of *O. diffusa* contains oldenlandosides, stigmasterol, ursolic acid, oleanolic acid, beta-sitosterol, p-coumaric acid and flavonoid glycosides. Ursolic acid inhibits growth and spread of various cancers such as cancers of lung, ovary, uterus, stomach, liver, colon, rectum and brain, malignant melanoma, malignant ascites, lymphosarcoma, and leukaemia. Ursolic acid works by a typical cytotoxic effect on cancer cells and by inducing apoptosis[1].

Panax ginseng Mey./_P. schinseng_ Nees (Asiatic or Chinese Ginseng; Family: Araliaceae)

Ginsenosides (panaxadiol and panaxatriol saponins), isolated from *P. ginseng,* inhibits the growth and spread of various cancers such as cancers of breast, ovary, lung, prostate and colon, renal cell carcinoma, malignant melanoma, malignant lymphoma, and leukaemia. Panaxadiol ginsenosides (Rb1, Rb2, Rc, Rd, Rg3, Rh2) and Panaxatriols ginsenosides (Re, Rf, Rg1, Rg2, Rhi) have both preventive and therapeutic roles in cancer treatment. Ginsenosides possess strong anticancer activity against lung cancer and also prevent lung metastasis by blocking angiogenesis.

Compound K (a metabolite of ginsenosides) inhibits growth and spread of chemo-resistant lung cancer. Ginsenosides Rc, Rd, Rg1 and Re overcome (reverse) P-glycoprotein mediated multi-drug resistance to chemotherapy. Ginsenoside Rf helps in reducing doses of morphine in terminally ill cancer patients. Polysaccharides of *P. ginseng* possess strong immunoenhancing and anticancer activities against many cancers, particularly lung cancer. These polysaccharides also reduce side effects of chemotherapy and radiotherapy. *P. ginseng* also possesses antistress, hepatoprotective, haemopoietic, immunoenhancing, antioxidant, radioprotective, chemoprotective and antiinflammatory properties. It inhibits proliferation and seeding (metastases) in various cancers by inducing cell differentiation and apoptosis. It is effective in both hormone-responsive and hormone-refractory prostate and breast cancers.

***Plumbago zeylanica* Linn. (Chitarak/Chitra; Family: Plumbaginaceae)**

Plumbagin, isolated from *P. zeylanica* root inhibits growth and spread of breast cancer, liver cancer, fibrosarcoma, malignant ascites and leukaemia by inhibiting cancer cell proliferation. *P. zeylanica* also possesses strong antioxidant, hepatoprotective, neuroprotective and immunoenhancing properties.

***Podophyllum hexandrum* Royle/*P. emodi* Wall. ex Hook. f. and Thoms./ *P. peltatum* (Papra/Indian Podophyllum/Himalayan May Apple; Family: Berberidaceae)**

Podophyllotoxin and podophyllin (lignans), isolated from *P. hexandrum* inhibit growth and spread of various cancers, including that of breast, ovary, lung, liver, urinary bladder, testis and brain, neuroblastoma, Hodgkin's disease, non-Hodgkin's lymphoma, and leukaemia. Podophyllotoxin is the most active among all the natural anticancer compounds. *P. hexandrum* also possesses potent radioprotective and haemopoietic properties[2,7].

***Prunella vulgaris* Linn./*Brunella vulgaris* Linn. (Dharu; Family: Labiatae/ Lamiaceae)**

Ursolic acid and oleanolic acid, isolated from *P. vulgaris*, inhibit growth and spread of various cancers such as cancers of the breast, cervix, lung, oral cavity, oesophagus, stomach, colon and thyroid, malignant lymphoma, intracranial tumours and leukaemia. This herb also possesses immunoenhancing, hepatoprotective, antioxidant, anti-HIV and anti-Herpes properties. It has normoblastic effect on the bone marrow.

Psoralea corylifolia **Linn. (Babchi; Family: Papilionaceae/Fabaceae)**

Bavachinin, corylfolinin and psoralen, isolated from *P. corylifolia*, possess strong anticancer activity against lung cancer, liver cancer, osteosarcoma, fibrosarcoma, malignant ascites and leukaemia. Psoralen enhances immunity of the body by stimulating natural killer cell activity. Psoralidin isolated from this herb inhibits growth and spread of stomach and prostate cancers by inhibiting G2/M phase of cell cycle. Psoralidin induces apoptosis in both androgen-responsive and androgen-refractory prostate cancers. *P. corylifolia* also possesses strong antioxidant, immunomoenhancing and hepatoprotective properties.

Punica granatum **Linn. (Anar/Pomegranate; Family: Punicaceae)**

Fruit (Anar) of *P. granatum* contains alkaloids, anthocyanidines and vitamin C. *P. granatum* acted against solid tumour and ascites tumour in albino mice[2].

Rubia cordifolia **Linn. sensu Hook. f. (Manjit/Majith/Rosemary; Family: Rubiaceae)**

Root of *R. cordifolia* contains rubidianin, rubiadin, rosemary acids (*viz.*, RA-7, RA-700, RC-1 and RC-18), carnosic acid, purpurin, pseudopurpurin, alizarin and xanthopurpurin (purpuroxanthin). *R. cordifolia* root inhibits growth and spread of breast, ovary, cervix, colon and lung cancers, malignant ascites, malignant lymphoma, malignant melanoma (B16 melanoma), P388 cells, L1210 cells, sarcoma, and leukaemia. Rubiadin also possesses hepatoprotective activity[2,7].

Saussurea lappa **C.B. Clarke (Kut/Kur/Kuth/Costus; Family: Compositae/ Asteraceae)**

Sesquiterpenes and costunolide dehydrocostuslactone, isolated from *S. lappa*, inhibit growth and spread of breast cancer. Cynaropicrin from *S. lappa* possesses strong anticancer activity against malignant lymphoma and leukaemia. Costunolide also inhibits growth and spread of intestinal cancer. Mokkolactone content of this herb induces apoptosis in leukaemic cells. Shikokiols obtained from *S. lappa* exhibit anticancer activity against cancers of ovary, lung, colon and central nervous system. *S. lappa* inhibits growth and spread of cancers by arresting cancer cell division in G2 phase and inducing apoptosis.

***Solanum nigrum* Linn. (Makoi/Kakmachi/Vayasi/Black nightshade; Family: *Solanaceae*)**

Flavonoids (*e.g.*, quercetin) and alkaloids (*viz.*, solasodine, solanine and solamargine) are the main phytoconstituents of *S. nigrum* whole palnt or fruit which have been reported to act in various tumours. Solamargine and solasonine inhibit growth and spread of various cancers, including breast, liver, lung and cyst cancers, choriocarcinoma or chorioadenoma, and leukaemia. Solanine and solamargine have very strong anticancer actions against murine tumours. Steroidal glycosides (spirostane, furostane, spirosolane and pregnane), isolated from *S. nigrum,* inhibit growth and spread of colon cancer and pheochromocytoma. Glycoproteins obtained from *S. nigrum* have antiproliferative and apoptotic effects on colon and breast cancers. Polysaccharides of this herb have significant inhibitory effect on growth of cervical cancer. *S. nigrum* inhibits growth and spread of liver cancer by two distinct anticancer activities, *i.e.*, apoptosis (programmed cell death) and autophagy (autophagocytosis). Higher doses of *S. nigrum* induce apoptotic cell death, while lower doses lead to autophagocytic death of cancer cells. Lunasin, isolated from *S. nigrum* is a cancer-preventive peptide. *S. nigrum* and *S. lyrati* inhibit growth and spread of stomach cancer, sarcomas, malignant ascites and leukaemia. *S. nigrum* leaves extract has inhibitory effect against S 180, V 14 and Ec tumour models[2].

***Tinospora cordifolia* (Willd.) Miers ex Hook. f. and Thoms. (Giloe/ Amrita/Gulancha/Gulbel//Tinospora; Family: Menispermaceae)**

Stem bark and fruit of *T. cordifolia* contain berberine, tinosporine, giloin and giloinin. Sesquiterpenes and diterpenes, isolated from this herb inhibit growth and spread of various cancers, including cancers of lung, cervix and throat, and malignant ascites. Polysaccharide fraction of *T. cordifolia* inhibits lung metastasis. Arabinogalactan, syringine, cordiol, cordioside, cordifoliosides (A and B) obtained from *T. cordifolia* possesses significant immunoenhancing activity. *T. cordifolia* reduces side effects of radiotherapy and chemotherapy. This herb also possesses antioxidant, neuroprotective, hepatoprotective, antistress, antiulcer, antiasthmatic, antidiabetic (or hypoglycaemic and hypolipidaemic) and antipyretic activities[7].

Viscum album Linn./_V. costatum_ Gamble (Banda/Ban/European Mistletoe; Family: Loranthaceae/Viscaceae)

V. album whole plant contains lectin alkaloids, acetylcholine, proprionyl choline, lupeol, viscotoxin, flavonoid and sterol A. Lectins (*e.g.*, viscumin), polypeptides (viscotoxins) and phenolic compounds (*e.g.*, digallic acid), isolated from *V. album* inhibit growth and spread of various cancers, including that of breast, cervix, ovary, lung, stomach, colon, rectum, kidney, urinary bladder and testis, malignant melanoma, sarcoma, fibrosarcoma, malignant ascites, lung metastasis, and leukaemia by inducing apoptosis and antiangiogenesis activity. Lectins possess both anticancer and immunostimulating activities. Lectin-II induces apoptosis in cancer cells via activation of caspase-3 cascades. Lectin alkaloids also cause lectin 11-induced apoptosis and inhibition of telomerase via mitochondrial controlled pathway independent of p53. Hexamethylene bioacetamide obtained from *V. album* causes p53-dependent apoptosis and induction with telomerase. Viscumin, responsible for most of the biological activities of *V. album*, acts by bringing together immune system effector cells and cancer cells[2,7].

Withania somnifera Dunal (Ashwagandha/Asgandh/Punir; Family: Solanaceae)

The majority of the phytoconstituents of *W. somnifera* root are withanolides (steroidal lactones with ergostane skeleton) and alkaloids. They include withanone, withaferin A, and several other withanolides and withasonidienone. Much of the pharmacological activity of *W. somnifera* has been attributed to two main withanolides, withaferin A and withanolide D. Apart from these, *W. somnifera* root also contains withaniol, acylsteryl glucosides, starch, glycosides, reducing sugar, resins, saponins, fixed oils, hantreacotane, ducitol, anthraquinones, proteins, amino acids (*e.g.*, aspartic acid, proline, tyrosine, alanine, glycine, glutamic acid, cystine and tryptophan) and high amount of iron, etc. Withaferin A and withanolide D have antioxidant, anticancer and immunoenhancing activities, and act against various cancers. Withanolides are similar to ginsenosides (the active principles of *P. ginseng*) in both structure and activity. Withanolides (including withaferin A, sitoindoside IX, physagulin D, withanoside IV and viscosalactone B) inhibit growth and spread of various cancers such as cancers of breast, lung, colon and central nervous system due to their antiproliferative and antiangiogenic properties. Withaferin A (the most important withanolide) inhibits growth and spread of various cancers, including that of breast, cervix, colon, prostate, nasopharynx and larynx, malignant ascites, and sarcoma by inducing apoptosis. Withaferin A is effective in both androgen-responsive and androgen-refractory prostate cancers. Sitoindosides VII-X and withaferin A have strong antioxidant, antistress, immunomodulatory, antiinflammatory and antiaging properties. Withanolide D inhibits the metastatic colony formation in malignant lung melanoma. Ashwagandhanolide, a new dimeric withanolide, isolated from *W. somnifera,* inhibits growth and spread of breast, stomach, colon, lung and central nervous system cancers. *W. somnifera* reduced the cancer cell proliferation and increased the overall survival time. It enhanced the effectiveness of radiation therapy, and reduced the side effects of radiotherapy and chemotherapy. Given its broad spectrum of cytotoxic and anticancer activity, *W. somnifera* presents itself as a novel therapy for cancer. *W. somnifera* root also possesses other medicinal properties like haemopoietic, neuroprotective, anticonvulsant, hypoglycaemic and hypolipidaemic properties[2,7,17].

***Zingiber officinale* Rosc. (Adrak/Ada/Ginger; Family: *Zingiberaceae*)**

Gingerols, isolated from *Z. officinale* rhizome inhibit growth and spread of various cancers, including that of ovary, cervix, colon, rectum, liver, urinary bladder and oral cavity, neuroblastoma, and leukaemia by inducing apoptosis. The most active individual component, 6-shogaol, isolated from *Z. officinale,* inhibits growth and spread of many cancers, particularly the ovarian cancer by blocking formation of new blood vessels, and by inducing apoptosis and autophagy. It is effective even in chemotherapy-resistant ovarian cancer. *Z. officinale* reduces side effects of chemotherapy and radiotherapy. It also possesses antioxidant, antimutagenic and antiinflammatory activities[2].

References

1. Madhuri S, Pandey Govind. Some anticancer medicinal plants of foreign origin. Curr Sci 2009; 96(6):779-783.
2. Pandey Govind, Madhuri S. Some anticancer agents from plant origin. Pl Arch 2008; 8(2): 527-532.

3. Chopra RN, Nayar SL, Chopra IC. Glossary of Indian Medicinal Plants. New Delhi: Council of Scientific and Industrial Research; 2002.
4. CSIR. The Useful Plants of India. New Delhi, India: Council of Scientific and Industrial Research; 1986.
5. IDMA. Indian Herbal Pharmacopoeia. Mumbai, India; 2002.
6. Prajapati ND, Purohit SS, Sharma AK, Kumar T. A Hand Book of Medicinal Plants. India: Agrobios; 2003.
7. Madhuri S, Pandey Govind. Ethnomedicinal plants for prevention and treatment of tumours. Int J Green Pharm 2009; 3(1):2-5.
8. Pandey Govind. Antioxidant vegetables act against cancer and other diseases. Int J Pharmaceu Stud Res 2011; 2(1):32-38.
9. Pandey Govind, Madhuri S. Some medicinal plants as natural anticancer agents. PHCOG REV 2009; 3(6):259-63.
10. Pandey Govind, Madhuri S. Therapeutic approach to cancer by vegetables with antioxidant activity. Int Res J Pharm 2011; 2(1):10-13.
11. Gupta VK, Sharma, SK. Plants as natural antioxidants. Nat Prod Rad 2006; 5(4):326-334.
12. Polidori MC. Antioxidant micronutrients in the prevention of age-related diseases. J Postgrad Med 2003; 49:229-235.
13. Pandey Govind. Active principles and median lethal dose of *Curcuma longa* Linn. Int Res J Pharm 2011; 2(5):239-241.
14. Pandey Govind, Pandey SP. Phytochemical and toxicity study of *Emblica officinalis* (Amla). Int Res J Pharm 2011; 2(3):270-272.
15. Hirazumi A, Furusawa E. An immunomodulatory polysaccharide-rich substance from the fruit juice of *Morinda citrifolia* (Noni) with antitumour activity. Phytother Res 1999; 13:380-387.
16. Pandey Govind, Madhuri S. Pharmacological activities of *Ocimum sanctum* (Tulsi): A review. Int J Pharmaceu Sci Rev Res 2010; 5(1):61-66.
17. Madhuri S, Pandey Govind. Anticancer activity of *Withania somnifera* Dunal (Ashwagandha). Indian Drugs 2009; 46(8):603-609.

Phytotherapy of Malnutritional Cancers in Animals

Govind Pandey[1] * and Y.P. Sahni[2]

[1]Ex-Professor/Principal Scientist and Head of Pharmacology (Pharmacy); presently Officer-In-Charge of Rinder Pest (Animal Husbandry Deptt., Govt. of MP), Jabalpur Division, Jabalpur

[2]Professor and Head of Pharmacology and Toxicology, College of Veterinary Science and AH, Madhya Pradesh Pashu Chikitsa Vigyan Vishwavidyalaya, Jabalpur, M.P., India

ABSTRACT

In the so called 'Third World', where poverty is common, millions of poor people are engaged in hard labours. Millions and millions of animals like horses, donkeys, bullocks and he-buffaloes, etc. work day in day out in cart or for other carriage. All of them don't get good food and don't get enough to drink. They have to work in the burning sun, on busy and dusty roads, without one moment of rest. Next to that, some individuals don't even know that 'What is pain'. A lot of poor people and different species of animals have to work till they drop dead, and that often happens in the streets. They are day-to-day exposed to numerous environmental hazards and disasters, including xenobiotics, contaminants, radiations, genetic factors, etc. All

Accepted on: 18-02-2011; Finalized on: 28-04-2011.

[Paper published in: International Journal of Pharmaceutical Sciences Review and Research, 2011; 8(1): 14-20].

*Corresponding author's E-mail: drgovindpandey@rediffmail.com

these result in 'malnutrition' which may cause 'cancer', *i.e.*, "malnutritional cancer". Overall, the poverty and decreased dietary intake of good quality are the most important causes of malnutritional cancer. From ancient times we, in India have endowed food with magical qualities. The episodes in the history of man's battle against disease clearly illustrate the fact that food can cure many diseases. Thus, food or 'phytotherapy' can be the best remedy for malnutritional cancer. The medicinal plants, including fruits and vegetables have versatile immunomodulatory and anticancer activities. Consumption of large amounts of fruits and vegetables can prevent the development of cancer. Many doctors recommend that people wishing to reduce risk of cancer eat several pieces of fruits and several portions of vegetables every day. This review article, therefore, explore out the certain phytotherapies (remedies) against malnutritional cancers occurring in the animals.

Keywords: *Animals, Malnutritional cancers, Phytotherapy (Edible plants), Vegetables and fruits.*

Introduction

Cancer, a disease of human and other multicellular animals, is formed as a result of abnormal, excessive, autonomous and purposeless proliferation of cells. Cancer is the largest (or second leading) cause of mortality in the world, is and caused by a complex, poorly understood interplay of genetic and environmental factors. Reports show that several environmental factors, including air, water and industrial pollutants, environmental chemicals and radiations, etc. may cause various cancers[1–2]. The incidence of cancer is increasing worldwide, and millions and millions of humans and animals are died every year. The animals like horses, donkeys, bullocks and he-buffaloes, etc. work day in day out in cart or for other carriage. All of them don't get good food and don't get enough to drink. They have to work in the burning sun, on busy and dusty roads, without one moment of rest. A lot of animals have to work till they drop dead, and that often happens in the streets. They are day-to-day exposed to numerous environmental hazards and disasters, including xenobiotics, contaminants, radiations, genetic factors, etc. All these result in 'malnutrition' which may cause "malnutritional cancer". Overall, the poverty and decreased dietary intake of good quality are the most important causes of malnutritional cancer. Now, the cancer is a burning problem in this westernized 'Ultra-Modern Era' even in our country.

Eating too little or too much will result in poor health. The food eaten must not only be nutritious but it must also be complete and clean, otherwise the individual eating the food would get ill even the food is nutritious. So, if the right food is not consumed in right amounts, it results in "malnutrition" (mal-noo-trish-un) which may lead to either 'undernutrition' or 'overnutrition'[1]. Malnutrition occurs when we do not get enough calories or nutrients and hence, it may cause several diseases, including cancer. Malnutrition acts as carcinogen in many ways to produce cancer in human beings and animals, and so the aetiological role of malnutrition preceding clinical cancer has been firmly established[1,3]. In India, malnutrition is the single largest cause of the mortality and morbidity not only in infants, children and working or pregnant women, but also in livestock animals. The single major factor responsible for the wide prevalence of malnutrition in India is the poverty[1]. Like many countries, India is also facing a great problem of malnutrition. Thus, similar to the human beings, the animals (*viz.*, horse, cattle, buffalo, sheep, goat, pig, dog, poultry, and also wild animals, etc.) suffering from malnutrition need special attention. Those in moderate and mild degree of malnutrition require additional food in order to bring their weight to normal; whereas, those suffering from a severe degree of malnutrition need special attention by the owner as well as veterinarian because such animals usually suffer from diarrhoea, respiratory infections and other diseases. It has been clearly stated[1,4] that malnutrition predisposes to infection, or infections lead to malnutrition. The body requires energy to carry out the different activities, and proteins are necessary for growth and repair. Deficiency of these two basic nutrients in our body leads to "protein energy malnutrition" (PEM). This disorder has very serious consequences, leading to cancer and/or death.

From ancient times we, in India, have endowed food with magical qualities. In the Vedic times, food was associated with divine attributes. This tradition was prevalent in other societies as well- ancient Egypt, for example. Ancient people discovered the healing powers of food- perhaps initially by accident. Later, these patterns became well established and were described in medical and veterinary texts. It is quite evident that the diseases which can be cured by food or nutrient concentrates are caused by deficiency of particular nutrients. Thus the malnutritional disorders can be cured by giving food sources rich in the nutrient that is otherwise lacking in the diet, and there is no doubt that food can help to control several ailments[1]. The treatment of malnutrition may include increasing the calories and nutrients in diet. If malnutrition results in cancer, it is very difficult to treat, but some improvements can be brought about by proper foods and drugs[1,5]. Modern (allopathic) drugs, being a major treatment modality for the treatment of diseases, exhibit severe toxicity on normal tissues, so the 'phytotherapy' is needed. The edible plants serve as a food and medicinal sources, which maintain the health and vitality of the individuals with no ill effect. India is the largest producer of medicinal plants and is rightly called the "Botanical garden of the World". In the indigenous system of medicine, numerous medicinal plants and their formulations are now being used for the treatment and control of various diseases, including cancer. More than 50 per cent of all modern drugs in clinical use are of natural products, many of which have the ability to include apoptosis in various tumour cells. Medicinal plants are known to have good immunomodulatory properties. They act by stimulating both non-specific and specific immunity. They may promote host resistance against infection by re-stabilizing body equilibrium and conditioning the body tissues[1,7–8]. Some medicinal plants and their products, including vegetables, fruits and crops play an important role for cancer prevention. Consumption of large amount of vegetables and fruits may prevent from the cancer. Many doctors recommend that people wishing to reduce their risk of cancer eat several pieces of fruits and several portions of vegetables every day. Many plant-derived products exhibit potent anticancer activity against several cancers. Thus, it is distinct that medicinal plants, including edible plants (fruits and vegetables) elicit immunomodulatory and anticancer activities[8–11].

With the above backgrounds, an attempt has been taken through this review article to elucidate on the malnutritional cancers occurring in animals, and their remedial approach by phytotherapy (edible plants) with special reference to vegetables and fruits.

Malnutritional Cancers in Animals

If an individual eats the right kind of foods in the required amounts, that individual will keep good health provided no other factors intervene. On the other hand, a poor eating pattern or eating too little or too much will result in poor health. These are both facets of "malnutrition". When the diet supplies too little of one or more nutrients, a kind of malnutrition called 'undernutrition' occurs. When the diet provides too much of one or more nutrients, another form of malnutrition called 'overnutrition' results. Malnutrition most often refers to undernutrition resulting from an inadequate consumption, poor absorption, or excessive loss of nutrients, but the term can also encompass overnutrition resulting from excessive intake of specific nutrients. An individual will experience malnutrition if the appropriate amount of, or quality of nutrients comprising a healthy diet are not consumed. An extended period of malnutrition can result in starvation or many other diseases such as scurvy and cancer. The individual suffering from malnutrition almost always have infections such as diarrhoea and respiratory infections. Just as infections are common in cases of severe malnutrition, infections like diarrhoea and whooping cough can also cause malnutrition. Thus, malnutrition can increase the risk of infections and infections can, in turn, lead to malnutrition. This inter-relationship and synergistic effect of malnutrition and infections often lead to a high incidence of cancer and/or deaths. Malnutrition leads to infection by reducing the diseases-fighting capacity of the body[1]. Many environmental or pollution factors, including poverty contribute to eating disorder or malnutrition.

In animals, PEM may result due to deficiency of energy and protein. This disorder is of three types, namely: (a) Marasmus (deficiency of both carbohydrate and protein); (b) Kwashiorkor (deficiency of only protein); and (c) an intermediate state of marasmus and kwashiorkor[1,4,12]. 'Marasmus' is characterized by very low body weight and weight loss due to growth failure, stunting, loss of subcutaneous fat, gross muscle wasting, emaciation and failure to thrive. In addition, watery diarrhoea is associated often with dehydration, irritation (does'nt allow to touch), decreased responsiveness, behavioural changes, no oedema and wrinkled dry skin dry. Other deficiencies particularly, vitamin A deficiency may occur. 'Kwashiorkor' is characterized by oedema of whole body (especially belly), wasted muscles and retarded growth. Other signs and symptoms include psychomotor changes, hepatomegaly (enlarged, fatty liver), irritability, skin changes (flaky paint appearance and mosaic skin appearance), vitamins A and B deficiencies and associated diseases such as watery diarrhoea, severe respiratory infection (cough), etc. In intermediate type of PEM, the features of both marasmus and kwashiorkor PEM are seen. Associated symptoms like vitamin C deficiency, zinc deficiency, atrophy of the papillae on the tongue, angular stomatitis, xerophthalmia, etc. may occur[12]. PEM can result in marked hypoalbuminemia, anaemia, oedema, muscle atrophy, delayed wound healing, impaired immunocompetence, weight loss, reduced basal metabolism, depletion of subcutaneous fat and tissue, bradycardia and hypothermia. The patient with PEM is typically emaciated, elderly and chronically ill. Thus, malnutrition places stressed patients at a greatly increased risk for morbidity, leading to cancer and mortality. More over, there may be a longer recovery period, impaired host defenses and sepsis, impaired wound healing, anaemia, impaired gastrointestinal tract function, muscle atrophy, impaired cardiac function, impaired respiratory function, reduced renal function, brain dysfunction, delayed bone callus formation and atrophic skin[1,4].

Cancer is primarily an environmental disease. In animals, the common environmental factors leading to cancer may be bad feed or diet, infections, radiation, hard working and environmental pollutants. These environmental factors cause abnormalities in the genetic material of cells. Cell reproduction is an extremely complex process, which is normally tightly regulated by several classes of genes, including oncogenes and tumuor suppressor genes. Hereditary or acquired abnormalities in these regulatory genes can lead to developmemt of cancer[12]. Abnormal diets such as fad or limited diet and recent decrease in food intake usually cause malnutrition[4]. There is an increased risk of malnutrition associated with chronic diseases such as cancer, and diseases of intestinal tract, kidney and liver. Patients with these chronic diseases, especially cancer may lose weight rapidly. Studies show that malnutrition is closely linked to the income level of the owner of animal. The animals given with high drug doses are also at increased risk of malnutritional cancer. They tend to maintain inadequate diets for long periods and their ability to absorb nutrients is impaired by the drug's effect on body tissues, particularly the liver, pancreas and brain. Malnutrition usually develops in stages over a long period of time. Some early signs and symptoms may include: irritable (bad mood) and tired, slower growth than normal, or no growth, and weight loss. Bone or joint pain, weak muscles, bloated abdomen, swelling in other parts of the body, change in the colour of skin and hair, hair loss, loss of appetite, slow wound healing, and easily get infections are noticed. All these signs and symptoms are correlated with those of many cancers. There is an increased risk of malnutrition associated with chronic diseases, *e.g.*, diseases of intestinal tract, kidney and liver, and cancer. Animals with these chronic diseases, especially cancer may lose weight rapidly and become susceptible to undernourishment because they cannot absorb valuable vitamins, calories and iron[1].

In relation to cancer, malnutrition is the carcinogenic effect produced by nutritional variables through multiple endogenous involvements. Nutritional factors act as primary effectors in four situations: carcinogens in food articles, affected bioavailability of nutrients, non-nutritive dietary items and harmful contaminants. Nutritional carcinogenesis may occur due to ingestion of toxins, malnutrition, non-bioavailability of the micronutrients and inactivation of the metabolic enzymes (*i.e.*, mixed-function oxidases) present in the liver. Thus the aetiological role of malnutrition preceding clinical cancer has been firmly established[1,3]. There are certain environmental substances (contaminants or hazards) which can cause malnutritional cancer both in humans and animals. Several reports indicate that environmental and food contaminants, including xenobiotics (environmental or foreign chemicals) such as oestrogenic endocrine disruptors (xenoestrogens), certain therapeutic agents (*e.g.*, antibiotics, antihypertensive drugs, etc.) may cause various types of cancer. The 'endocrine disruptors' influence the normal functions of oestrogens, thereby cause the cancers of several organs. Chemicals, industrial wastes, pesticides, oral contraceptives (OCs) or oestrogens, detergents, food additives and plastics, all are sources of environmental toxins and endocrine disruptors. The potential endocrine disrupters such as antibiotics, hormones, plasticizers and non-ionic surfactants are becoming the priority pollutants. Most of these have been shown as carcinogenic, or they may contaminate the diet or food, thereby causing malnutrition. Many factors have a major effect on increasing the rates of oral, colon, lung and mammary gland cancers. Some of these factors include increased infections, more use of pesticides, low consumption of fruits and vegetables, high industrial pollution, more exposure to sun, decreased physical activity, etc. Almost every fruit, vegetable, etc. contains natural carcinogenic pesticides. Endocrine disruptors are xenobiotic (environmental/ foreign) chemicals that adversely interfere with the natural functions of hormones[13–14].

Malnutrition is very common in sick animals and is thought to affect up to 50 per cent of hospitalized dogs and cats. Within days malnutrition can begin to have deleterious effects on immune function, healing and, most importantly, quality of life. While human patients will follow their doctor's instruction to eat even though they don't feel well or have a good appetite, it is impossible to convince a sick pet to eat if they don't feel like it[15]. Due to malnutrition, different types of cancer may develop in dogs. Some symptoms of dog cancer are rather indistinct and, especially at the onset of the disease, it is difficult to tell whether a dog showing some cancer-like symptoms is in fact suffering from canine cancer or some other illnesses. Some cancers (*e.g.*, lymphomas, mast cell tumors, fibrosarcomas) can arise as lumps and bumps on the dog's skin. Very often, benign growths such as warts, fatty tumours, cysts, etc. can arise in a dog's skin as well. Usually, any lumps and bumps that appear, then decrease in size or even disappear, and finally reappear in a larger size are more likely to be cancer than benign lumps. While coughing and sneezing are common symptoms of dog cancer, there are also quite a few other dog illnesses that have the same symptoms. Chronic disorders of the gastrointestinal tract such as vomiting and diarrhoea can be symptoms of dog cancer of the gastrointestinal tract, or cancer anywhere else. Appetite loss is also one of those symptoms of dog cancer that is indistinct, as many other diseases (*e.g.*, dental or gum diseases, inflammatory bowel disease, heart disease, kidney disease, etc.) can cause appetite loss in dogs as well. Also, older dogs tend to eat less. Chronic skin diseases such as skin sores, itch and lesions, and wounds that do not heal, can be indicative of canine skin cancer. Blood in urine or urinary incontinence can be caused by some types of dog cancer (*e.g.*, bladder cancer), but they can also be caused by urinary tract infections, bladder stones, kidney disease, diabetes, Cushing's disease, etc. While cancer can cause a lot of pain to a dog, so can many other diseases, such as arthritis, ear infections, dental or gum diseases, etc. Bone cancer in dogs can cause lameness. However, lameness can also be caused by arthritis, hip dysplasia, or some form of injuries. Perhaps lethargy and weakness are one of the most vague symptoms of all diseases. A dog can be lethargic due to a various reasons, *e.g.*, pain from arthritis, anemia, fever caused by infections, etc. Oral cancer in dogs can cause bad breath, but perhaps more frequently, bad breath is caused by indigestion, dental/gum diseases, liver disease and kidney disease. Brain tumours in dogs can cause seizure, but it can also be seen in thyroid problems, kidney disease, liver disease, hypoglycaemia and poisoning. Many of the above symptoms of dog cancer are not unique to cancer. That is why very often cancer is not diagnosed until it is at a later stage. Finally, whenever a dog has a disorder that does not respond well to treatment in one to two months, and is showing some symptoms of dog cancer, then the possibility of cancer should be considered and a more thorough examination should be done[16].

The owners of the horses and donkeys are often very poor. The owner really needs the money, because he has to take care of his family. When the animal becomes sick, the owner is not able to make money because he doesn't have the horse or donkey to pull. In malnutrition, the horses grow out of shape. Because of all the heavy work, the horses can not develop physically; often legs and ankles become crooked. They have often a bad back also, since they have to carry a too heavy load and because it is loaded in a wrong manner. The animals have to work in the burning sun with terribly high temperature. The owners don't give them enough to drink so they dehydrate. Due to less consumption of water and food the animals are exhausted and overly tired. Through malnutrition, starvation, dehydration and exhaustion occur. In the countries like Netherland, Egypt, Jordan, Pakistan and India, the cars are expensive so horses pull the carriages that transport people; and horses and donkeys have to pull carts with heavy loads such as building materials. These animals are seen as 'motor' to do the heavy works. These countries are usually very hot and we cannot cope with that, but the horses and donkeys have a problem with it also. In World War I (1914-1918), the British army was present in Egypt. In the beginning of this war, about 20,000 horses and donkeys from the UK were brought ashore. When the war ended in 1918, the horses and donkeys were sold to Egyptian people so the poor animals were left there[17].

Edible Plants as Phytotherapy for Malnutritional Cancers

Treatment for malnutrition depends on the cause. Vitamin or mineral supplements are needed if the body is low in vitamins and minerals. Medicines may also be given if health problem appears due to malnutrition. Diet therapy helps patients to lead a full life. To prevent energy deficiency, carbohydrate-rich foods (*e.g.*, sugars, cereals, roots and tubers, fruits like banana, sapota and mango) and fat-rich foods (*e.g.*, nuts, oilseeds, vegetable oil, ghee, vanaspati and butter) should be consumed. If protein is lacking in the diet, pulses, nuts and oilseeds should be given. If vitamins and minerals are deficient in the diet, green leafy vegetables, nuts, pulses, etc. are to be eaten. In addition, fibre also plays role in the prevention of diseases like constipation and cancer of the colon. Some complex carbohydrates (*e.g.*, cellulose) and non-carbohydrate forms of fibre are available, which may be helpful in treating the colon cancer. In PEM, high energy foods in the case of mild to moderate diarrhoea, a high energy-high protein diet when tolerance improves for solid foods, and mineral and vitamin supplements, etc. are given. In chronic diseases like cancer in which tissue wasting takes place, a high caloric-high protein diet is the best[1].

Besides above, 'phytotherapy' plays an important role in the prevention and treatment of many cancers. It has been reported that the diets rich in plant foods can lower the risk of many chronic diseases, including cancer[18]. Similarly, the herbivorous animals suffering from malnutritional cancer must be given the feed containing sufficient quantity of fruits and vegetables. The studies strongly suggest that vegetables and fruits have strong protective effect against major diseases, including cancer. The protective action of fruits and vegetables against cancer has been attributed to the presence of antioxidants, especially antioxidant vitamins. Fruits and vegetables contain several phytochemicals which possess strong antioxidant activities. Thus the fruits and vegetables prevent from cancer and other diseases by protecting cells from damage caused by 'free radicals'-highly reactive oxygen compounds. Certain phytochemical antioxidants with anticancer activity include vitamins (A, C, E, K), carotenoids, terpenoids, polyphenols (ellagic acid, gallic acid, tannins), flavonoids (quercetin, anthocyanins, catechins, flavones, flavonones, isoflavones), enzymes (superoxide dismutase, catalase, glutathion peroxidase), minerals

(Cu, Mn, Se, Zn), polysaccharides, saponins, lignins and xanthones[6,9-10,19-20].

Some agricultural plants and their products such as vegetables, fruits and crops play an important role for the prevention of cancer. Many naturally occurring substances present in the diet/food have been identified as potential chemopreventive agents. Various agriculture plants and their products have been used traditionally for the treatment of various livestock diseases since times immemorial. The dependence of rural mass on the plant-based medicines for curing animals is mainly because of the limited access to the modern (allopathic) medicine system, cheaper and easy availability, and the simplicity of their applications. The herbal drugs also do not possess toxicity or have very less side effects[21]. The edible plants are used not only for food but also employed to cure different ailments of livestock. The medicinal uses of such plants may help veterinarians, medicos and agriculture/bio scientists or researchers in discovering new drugs against various diseases of human and animals[22].

Many dietary agricultural plants and their products have been reported to possess versatile anticancer and immunomodulatory properties. They are commonly consumed by humans as well as animals as food matters, and prevent from malnutritional cancer and various other diseases. Some of the dietary plants which have been reported by various authors[9,11,18,21-26] as 'phytotherapy' against malnutrition and cancer are described below:

The bulb of *Allium cepa* (Piyaz) has been reported to contain diallyl disulphide (a sulphurous product), quercetin flavonoid, allicin, allin, amino acids and vitamins (folic acid, C and E). Diallyl disulphide inhibited the cancerous cells in stomach. Quercetin, because of its antioxidant properties has been found to treat lung cancer and some other cancers. The bulb of *Allium cepa* var. *aggregatum* (Shallot) contains allyl propyl disulphide (a sulphurous product), flavonoid, phenolic content, proteins and vitamins (folic acid and C). Allyl propyl disulphide, phenolic content and flavonoid inhibited the cancerous cells in liver, stomach and other organs. It also eliminated the toxins to inhibit and kill yhe cancer cells in liver. *Allium sativum* (Lasun) bulb contains sulphur compounds (diallyl sulphide, diallyl disulphide, allyl propyl disulphide), allicin, allin and amino acids. The sulphur compounds inhibit cell proliferation, modulate cell cycle activity and interfere with hormone action in cancer cells. Allicin inhibits the proliferation of cancerous cells of mammary gland, endometrium and colon. Lasun is also used in gastric and liver cancers. In animals, the bulbs also are given in malnutritional disorders of skin and lung, flatulence, and dyspepsia. *Annona squamosa* (Sitaphal) contains annoaine, higenamine and amino acids. Its leaves and seeds are externally applied over malnutritional wounds, boils and swelling in animals. The riped fruit (sitaphal) is rich of carbohydrate (energy), so it can be given to cure marasmus PEM. *Brassica campestris* (Sarson) seed oil contains glycerides of palmitic, stearic, oleic and linoleic acids. It is used in various tumours. In animals, the oil is also given in rheumatism, cancer and ulcer. *Brassica juncea* (*B. nigra*, Rai) plant contains essential oils, while its seed contains 2 antithiamines and flavonol glycosides. In animals, oil along with rock salt is used in the diseases of gum. Seeds and oil of rai are used in malnutritional rheumatism, paralysis and skin diseases. The sprouts (flowers) of *Brassica oleracea* var. *botrytis* (Phoolgobhi) contain cysteine, ascorbigen, sulphoxide, indole-3-carbinol, glucaric acid, sulphoraphane, glucosinolates, isothiocyanates and vitamins (A, C). These components possess anticancer activities against lung, stomach, colon, bladder, mammary and rectum cancers. Its leaves contain vitamin C, and are externally used in gout, rheumatism, skin diseases and blister in animals; while phoolgobhi is cardiotonic and antiinflammatory, and may be used in biliousness and urinary diseases. *Brassica oleracea* var. *capitata* (Pattagobhi, cabbage) contains glucaric acid, sulphoraphane, glucosinolates, isothiocyanates, cysteine, ascorbigen, sulphoxide, indole-3-carbinol, allyl isothionate and vitamins (A, B, C). These have anticancer activities against bladder, lung, stomach, colon, rectum and mammary cancers. *Brassica rapa* (Shaljam) leaves and roots contain ascorbigen and vitamins (A, C), and are used in various tumours. *Cajanus cajan* (Arhar) leaves contain cajanol, while roots contain cajanone and cajaquinone. Its seeds are quite rich in proteins, so the seeds (or cooked 'dal') can be given in marasmus and kwashiorkor PEM. The expressed juice of leaves, with salt is given in animals to cure malnutritional jaundice.

The fruits of *Capsicum annum* (Lalmirch) contain amino, ascorbic and folic acids, glutathione, and flavonoids. In animals, the fruit powder with oil cures foot diseases. Fruits are stimulant and stomachic, and are used in dyspepsia, neuralgia, rheumatism and diarrhoea. *Carica papaya* (Papita) fruit contains papin and cryptoflavin, while the seeds contain carpasemine. In animals, the seeds and fruits are given to cure hepatitis, diarrhoea and dysentery. The fruits are digestive, stomachic, carminative and diuretic too. *Cicer arietinum* (Chana) seed contains isoliquiritigenin, while the plant contains amino acids and vitamins (A, D, E). Its fresh leaves juice is given to livestock in constipation. The macerated seeds in water are used as tonic. The seeds are astringent, diuretic and antibilious, and are given in debility, scurvy and jaundice. Chana seeds (or cooked 'dal') are quite rich in proteins, so the seeds (or cooked 'dal') can be very useful in marasmus and kwashiorkor PEM diseases. *Citrus limon* (Baranibu) fruit contains flavonoid, flavone, limonoid, limonene, nobiletin, tangeretin and vitamin C. The flavonoid, tangeretin and nobiletin are potent inhibitors of tumor cell growth and can activate the detoxifying P-450 enzyme system. Limonoids inhibit tumour formation by stimulating the enzyme glutathione S-transferase (GST). The limonene (a terpenoid) also possesses anticancer activity. Nibu fruit is used for inhibition of mammary cancer cell proliferation and delaying of mammary tumorigenesis. It is also used in metastasis and leukaemia. The leaves of *C. limon* contain bergapten, citropten and limettin. The fruits are antiseptic, appetizing, astringent and stomachic, and are given in scurvy, rheumatism, dysentery, diarrhoea and liver disorders of animals. *Curcuma longa* (Haldi) rhizome contains curcumin, zingiberine and curcuminoids. Haldi is reported to act against colon, bladder and prostate cancers, intravesical tumour, fibrosarcoma, hepatocellular carcinoma, oesophagal carcinogenesis, leukaemia, stomach papilloma and solid tumour cell lines. In large animals, the rhizome powder is used in pain, swelling, wound, mastitis, tumour and liver diseases. *Daucus sativa* (Gajar) root contains pyruvate kinase, acyltransferase and alkane; while the seeds contain sitosterol and amino acids. The boiled roots are given to cattle to fatten them. The roots and seeds are stimulant and carminative, and are used in dropsy and kidney diseases. In animals, the decoction of roots and leaves of *Elephantopus scaber* (Gobhi) are given in dysuria, diarrhoea, dysentery, swelling or pain in stomach, ulcer and eczema. Its whole plant is astringent and cardiac tonic. *Emblica officinalis* (Amla) fruit contains ascorbic acid (vitamin C) and phyllembic acid. It inhibits cellular mutation and may prevent cancer. Its 18 compounds inhibit the growth of gastric and uterine cancer cells. Amla also inhibits the growth of *in vitro* breast cancer cells. *Glycine javanica* (Soybean) seed contains phytates, protease inhibitors, phytosterols, saponins, isoflavonoids and isoflavones. The soybean is contributing factor in the low incidence of breast, prostate, stomach, colon, rectum and lung cancers. Its isoflavonoids inhibit the growth of hormone-dependent and hormone-independent cancer cells in culture; however, its isoflavones inhibit the growth of mammary and prostate cancers. The soy seeds (or cooked 'dal') are quite rich in proteins, so the seeds (vegetables) may potentially treat both marasmus and kwashiorkor PEM. *Lens culinaris* (Masur) plant contains lenticin, tricetin and luteolin; while seed contains protein. In large animals, masur gruel is given in haematuria, and paste of seeds is applied on the wounds and broken horns. The seeds are laxative and given in constipation. Since the seeds of *L. culinaris* are rich in protein, these (as masur dal) can be used to cure both marasmus and kwashiorkor PEM diseases.

Lycopersicon esculentum (Tamatar) fruits contain lycopene, which acts as an antioxidant and inhibits the proliferation of cancer cells. Higher intake of tamatar is correlated with protection from cancer; the protective effect was strongest for cancers of prostate, lung and stomach but some protective effects also appeared for cancers of pancreas, colon, rectum, esophagus (throat), mouth, mammary gland and cervix. Tomato fruits also contain oxalic, citric and mallic acids, and vitamins (A, C); while its seeds and herbage contain solanine. Its pulp and juice are intestinal antiseptic, blood purifier, cholagogue, stimulant and digestive, and are given in pulmonary diseases and dyspepsia of animals. *Momordica charantia* (Karela) leaf, fruit and seed contain linoleic acid, momordin, palmitic acid, proteolytic enzymes and vitamins (A, B, C). These parts are used in colon, mammary and bladder carcinomas, lymphoma, leukaemia and other tumours. The fruits and seeds are carminative and stomachic, and are prescribed in rheumatism, gout and liver diseases of animals. The silver skin of *Oryza sativa* (Chaaval) contains oryzanin; while bran contains a glucoside nukain, which yields aglucone nukagenin. In animals, chaaval with dal is used as a perfect food in dysentery and can alone support life as it contains protein, carbohydrate, fat, vitamins and minerals. Thus, chaaval may cure both marasmus and kwashiorkor PEM diseases.

Decoction of grains is used internally as antiinflammatory for digestive tract.

Picrorhiza kurroa (Kutki) fruits/seeds contain kutkin and kurrin. In animals, the fruits are given in loss of appetite, dropsy, fever, and foot and mouth disease (FMD). The roots are bitter, stomachic and purgative. *Psidium guajava* (Amrud) plant contains leucocyanidin and quercetin, while the leaves contain eugenol. In animals, the decoction of root bark and leaves is given in chronic diarrhoea. Its root bark is astringent and fruits are laxative; while the leaves are astringent, and applied in wounds and ulcers. *Raphanus sativus* (Muli) root contains glucoside, enzymes and methyl mercaptan. In animals, the roots are used in dysentery, indigestion, urinary disorders and gastrointestinal pain. Its seeds are diuretic, laxative and carminative. *Sesamum orientale* Linn. (Til) seeds contain essential oil, proteins sesamin and sesamolin, so its seeds and oils can be given in PEM diseases of animals. *Solanum melongena* (Baingan) fruit contains vitamins A, B and C; while seed contains linoleic and oleic acids. In animals, the fruits and leaves are given in weakness, dysuria, liver disorders and ulcers. The seeds of *S. melongena* are stimulant. *Solanum tuberosum* (Alu) plant contains solanine, solasodine and lutein. In livestock, a portion of raw tuber (alu) is grated over the sore to cure leg ulcers. Alu is full of carbohydrate, it can be given in weakness, particularly in PEM. *Spinacia oleracea* (Palak) plant contains vitamins, iodine, lecithin, carotene and amino acids. The whole plant and leaves of *S. oleracea* are used in inflammation of lung, liver and bowel, fever and jaundice in animals. Its seeds and fruits are laxative, demulcent and diuretic. *Trachyspermum ammi* (Ajwain) fruits yield essential oil containing thymol. In animals, the fruits are given in weakness after pregnancy and low lactation. They are also employed in gastric disorders and diarrhoea, and have antiseptic, carminative and stimulant properties. *Trigonella foenumgraecum* (Methi) leaves and seeds contain choline, trigonelline, amino acids, proteins, vitamins and quercetin. These parts are used in various cancers. In animals, its leaves and seeds are used in rheumatism, cancer, intestinal inflammation, indigestion and bilious disorders. Its seeds are carminative and tonic too. *Triticum aestivum* (Gehun) grains contain ellagic, linolenic and oleic acids, oligosaccharides, fat, sterols, phytases, tocotrienols and vitamin E. Whole grains protect against lymphomas and cancers of pancreas, stomach, colon, rectum, breast, uterus, mouth, throat, liver and thyroid. In animals, the flour of wheat is given in weakness and gastric disorders. *Zea mays* (Makka) grains contain oxalic acid, carbohydrate, fat, protein and vitamins. The grains are astringent, nutritive and nourishing in humans and animals. *Zingiber officinale* rhizome (called adrak and the dried rhizome called as sonth) contains a volatile oil containing camphene, gingerol, zingiberene, borneol, cineol and proteins. It is used in various cancers. In livestock animals, the rhizomes are given with salt before feed to increase appetite and in weakness after pregnancy. They are also given in swelling, indigestion, flatulent colic, and are stimulant and carminative. Apart from the above edible plants (phytotherapy), some other dietary plants which have been reported to exhibit anticancer properties are *Amorphophallus companulatus* (Suran), *Avena sativa* (Oat), *Cajanus cajan* (Arhar), *Hordeum vulgare* (Jau), *Lens culinaris* (Masur), *Mentha arvensis* (Podina) and *Zea mays* (Makka)[9,24–26].

References

1. Pandey Govind. Role of malnutrition in causing cancer and its safety measures. Diploma in Nutrition and Health Education (DNHE) project report, Indira Gandhi National Open University, New Delhi, India: 2007.
2. Pandey Govind, Madhuri S. Oncogenic DNA and RNA viruses causing the cancer pathogenesis. Int J Pharmaceu Sci Rev Res 5(1): 2010, 120-123.
3. Greenberg DS. Washington report: The Cancer Institute on the Griddle. *N Engl J Med* 299(4): 1978, 207-208.
4. Gottschlich MM. The Science and Practice of Nutrition Support. Kendall/Hunt Publishing, Dubuque: 2001.
5. Parkin DM, Bray F, Ferlay J, Pisani P. Global cancer statistics 2002. Cancer J Clin 55: 2005, 74-108.
6. Madhuri S. Studies on oestrogen induced uterine and ovarian carcinogenesis and effect of ProImmu in rats. PhD thesis, Rani Durgavati Vishwa Vidyalaya, Jabalpur, MP, India: 2008.
7. Madhuri S, Pandey Govind. Some anticancer medicinal plants of foreign origin. Curr Sci 96(6): 2009, 779-783.

8. Pandey Govind, Madhuri S. Therapeutic approach to cancer by vegetables with antioxidant activity. Int Res J Pharm 2(1): 2011, 10-13.
9. Madhuri S, Pandey Govind. Some dietary agricultural plants with anticancer properties. Pl Arch 8(1): 2008, 13-16.
10. Polidori MC. Antioxidant micronutrients in the prevention of age-related diseases. J Postgrad Med 49: 2003, 229-235.
11. Vecchia CL, Tawani A. Fruits, vegetables and human cancer. Eur J Cancer 7: 1998, 3-8.
12. Protein Energy Malnutrition.htm, Website: 2011.
13. Mukherjee S, Bidhan CK, Ray S, Ray A. Environmental contaminants in pathogenesis of breast cancer. Indian J Exp Biol 44: 2006, 597-617.
14. Pandey Govind, Madhuri S. Cancer pathogenesis caused by xenoestrogens of environment and food contaminants: A review. J Chem Pharm Res 2(4): 2010, 687-695.
15. Animal Specialty Hospital, Rockledge Florida.htm, Website: 2011.
16. Natural-Dog-Health-Remedies.com, Website: 2011.
17. Animal World-Brook Hospital for Animals.htm, Website: 2011.
18. Heber D. Vegetables, fruits and phytoestrogens in the prevention of diseases. J Postgrad Med 50: 2004, 145-149.
19. GuptaVK, Sharma SK. Plants as natural antioxidants. Nat Prod Rad 5(4): 2006, 326-334.
20. Kathiresan K, Guanasekan P, Rammurthy N, Govidswami S. Anticancer activity of *Ocimum sanctum*. Pharmaceutical Biology 37(4): 1999, 285-290.
21. Jadeja BA, Odedra NK, Solanki KM, Baraiya NM. Indigenous animal healthcare practices in district Porbandar, Gujarat. Indian J Trad Knowledge 5(2): 2006, 253-258.
22. Pandey Govind, Madhuri S. Agriculture plants used against livestock diseases. Int J Plant Sci 3(2): 2008, 671-674.
23. Chopra RN, Nayar SL, Chopra IC. Glossary of Indian Medicinal Plants. Publications and Information Directorate, CSIR, New Delhi, India: 2002.

24. CSIR. The Useful Plants of India. New Delhi, India: 1986.

25. Kaushik P, Dhiman AK. Medicinal Plants and Raw Drugs of India: Bishen Singh Mahendra Pal Singh, Dehradun, Uttarakhand, India: 1999.

26. Prajapati ND, Purohit SS, Sharma AK, Kumar T. A Hand Book of Medicinal Plant, 1st Ed. Agrobios, India: 2003, p. 367.

Therapeutic Approach to Cancer by Vegetables with Antioxidant Activity

Govind Pandey[1] * and S. Madhuri[2]

[1]*Ex-Professor/Principal Scientist and Head of Pharmacology (Pharmacy); presently Officer-In-Charge of Rinder Pest (Animal Husbandry Department, Govt. of MP), Jabalpur Division, Jabalpur, M.P., India*

[2]*Guest Faculty, Department of Zoology, Govt. MH College of Home Sci. and Science for Women, Jabalpur, M.P., India*

**Officer-In-Charge of Rinder Pest (Animal Husbandry/Veterinary Department, Govt. of MP), Jabalpur Division, Jabalpur, MP, India; E-mail: drgovindpandey@rediffmail.com, drgovindpandey@yahoo.co.in*

ABSTRACT

Cancer is the second leading cause of human deaths in the world. However, the potential treatment of cancer is still under investigation. In fact, the plants may occupy a good place in the treatment of cancer with no ill effect. The medicinal plants and their products, particularly vegetables have antioxidant activity leading to anticancer effect. Thus, more than 80 per cent people in developing countries depend on traditional medicine or plants for their primary health needs. Plants used as vegetables prevent human from several diseases, including cancer. Many doctors recommend that people wish to reduce the risk of cancer must eat several portions of vegetables every day. The vegetables contain many phytochemicals having antioxidant activity. The antioxidants protect the cells from damage caused by 'free oxygen radicals'. The main phytochemicals which show antioxidant activity are vitamins, carotenoids,

---

Article Received on: 24/12/10 Revised on: 12/01/11 Approved for publication: 17/01/11.

[Paper published in: International Research Journal of Pharmacy, 2011; 2(1): 10-13].

terpenoids, flavonoids, polyphenols, saponins, enzymes and minerals. Hence, the present article gives a better therapeutic approach to cancer by the maximum use of antioxidant vegetables against different cancers.

Keywords: *Therapeutic approach to cancer, Medicinal plants, Antioxidant vegetables, Phytochemicals.*

Introduction

Cancer (malignant tumour) is a disease of human and other multicellular animals. It may be uncontrollable and incurable, and may occur at any time at any age in any part of the body. On the other hand, cancer is a frightful disease because the patient suffers pain, disfigurement and loss of many physiological processes. It is caused by a complex, poorly understood interplay of genetic and environmental factors. Being chronic in nature it is difficult to cure cancer as the cells proliferate to their maximum. At present, cancer is the second leading cause of death and claims over 6 millions people every year[1–2].

Although many cancers can be controlled, the chronicity of the disease and expenses on supportive treatment are the valid reasons of not getting the cancer cured. Recent researches revolve round the urgency to evolve suitable chemotherapy consistent with new discoveries for the treatment of cancer with no toxic effect on normal cells. Toxicological aspects of drug warrant serious consideration without which the therapeutic concept of clinical efficacy will remain misconceived. Chemotherapy being a major treatment modality used for the control of advanced stages of malignancies and as a prophylactic against possible metastasis, exhibits severe toxicity on normal tissues. The plant kingdom serves as food and medicinal sources, and thus maintains the health and vitality of human beings as well as animals without causing any toxicity. India is the largest producer of medicinal plants and is rightly called the "Botanical garden of the World". In the indigenous or traditional system of medicine, many medicinal plants and their preparations are now used for the treatment of different diseases, including cancer[1,3–4]. More than 50 per cent of all modern drugs in clinical use are of natural products, many of which have the ability to include apoptosis in various tumour cells of human origin[4–5]. According to WHO estimates, more than 80 per cent people in developing countries depend on traditional medicine or plants for their primary health needs. A recent survey shows that more than 60 per cent of cancer patients use medicinal plants at some point in their therapy[3–4,6]. Medicinal plants, including vegetables are known to have good immunomodulatory antioxidant activities, leading to anticancer effect. They act by stimulating both non-specific and specific immunity, and may promote the host resistance against infection by re-stabilizing body equilibrium and conditioning the body tissues[1,3]. Hence, the consumption of vegetables is widely accepted as lowering the risk of different types of cancer. Vegetables contain several phytochemicals having potent antioxidant activities. The antioxidant vegetables prevent from the cancer by protecting cells from damage caused by 'free radicals'- highly reactive oxygen compounds. Thus, consuming a diet rich in vegetables with antioxidant activity may protect from the occurrence of cancer.

Some Scientific Reports on the Role of Vegetables in Cancer

Some antioxidant vegetables, including spices and oils have been mentioned in Table 1, describing the vegetables in alphabetical order along with their Hindi/English names, parts used, main active components and specific antioxidant/anticancer activities.

It has been recommended[7] by many doctors that people wish to reduce the risk of cancer must eat several portions of vegetables every day. A study of the WHO on diet, nutrition and prevention of chronic diseases recommended that we daily consume at least 400 g of vegetables, including at least 30 g of pulses, nuts and seeds. People who eat much quantity of vegetables have about one-half the risk of cancer and less mortality from cancer[8]. It has been further stated[8] that one-third of all cancer deaths in the United States could be avoided through dietary modification, which includes an abundant intake of vegetables and fruits. A large body of scientific evidence indicates that an association exists between inadequate antioxidant status and increased risk for many diseases, including cancer. Antioxidant vegetables have beneficial effect against several diseases. Vegetables contain compounds such as sulphoraphane that induces GSH transferase, there by helping detoxifying many carcinogens. Increased consumption of vegetables can increase the plasma antioxidant capacity and is associated with the lower risk of cancer. Vegetables are most effective against those cancers that involve epithelial cells such as cancers of lung, esophagus, stomach, colon, pancreas and cervix. The protective effect of vegetables has also been observed for hormone related cancers. The alcoholic extracts of dhaniya, haldi, karela and adrak were tested for tumour inhibitory effect. Haldi and karela were most effective against metastatic prostate cancer cell lines. Vegetables with the highest anticancer activity are lasun, soybean, pattagobhi, gajar; with a modest level of cancer-protective activity are piyaz, nibu, haldi, phoolgobhi, tamatar; with a low level of anticancer activity is khira[7,10–11].

The main phytochemicals which show antioxidant activity are vitamins (A, C, E, K), carotenoids, terpenoids, polyphenols (ellagic acid, gallic acid, tannins), flavonoids (quercetin, anthocyanins, catechins, flavones, flavonones, isoflavones), enzymes (superoxide dismutase, catalase, glutathion peroxidase), minerals (Cu, Mn, Se, Zn), polysaccharides, saponins, lignins and xanthones. All of these prevent from the cancer[4,8,12–13].

Some phytochemicals which provide protection against cancer are: allyl sulphides present in lasun and piyaz; glucarates in nibu, began and alu; phytates, lignans, isoflavones and saponins in soybean; isothiocyanates in sarson, phoolgobhi and pattagobhi; and flavonoids, carotenoids and terpenoids in different vegetables. These phytochemicals block various hormone actions and metabolic pathways which are associated with the development of cancer[8]. Soybean is the contributing factor in the low incidence of breast, prostate, stomach, colon, rectum and lung cancers. Soybean seed contains isoflavonoids which inhibit the growth of hormone-dependent and hormone-independent cancer cells in culture. The isoflavones in soy inhibit the growth of human breast and prostate cancers[9]. Nibu due to its flavonoid, tangeretin and nobiletin contents can potentially inhibits the tumor cell growth and can activate the detoxifying cytochrome P-450 enzyme system. Limonoids present in nibu inhibit tumor formation by stimulating glutathione S-transferase (GST) enzyme. Limonene (a terpenoid) content of nibu also possesses anticancer activity. Nibu has been used to inhibit the breast cancer cell proliferation, delay mammary tumorigenesis, and cure metastasis and leukaemia cancers[8–9]. A variety of isoprenoid compounds have been found in the vegetables which show anticancer activities. These compounds include tocotrienols (related to tocopherols) and terpenoids (*e.g.*, limonene, geranoil, menthol and

Table 1: Some antioxidant vegetables useful against cancer/tumour.

Sl.No.	Botanical (with Hindi or English) Name	Part Used	Main Active Components	Specific Antioxidant/Anticancer Activity	References
1.	*Abelmoschus esculentus* (L.) Moench (Bhindi, Lady's finger)	Fruit, seed	Carotene, vitamins (B, C) and amino acids.	Seeds inhibit cancer growth.	14-15
2.	*Allium cepa* Linn. (Piyaz, onion)	Bulb	Diallyl disulphide, allicin, allin, quercetin antioxidant flavonoid and vitamins (C, E).	Diallyl disulphide inhibits stomach cancer and quercetin may cure lung and other cancers.	7-8,10, 14-15
3.	*Allium sativum* Linn. (Lasun, garlic)	Bulb	Sulphur compounds (diallyl sulphide, diallyl disulphide, allyl propyl disulphide) and allicin.	Allicin inhibits cancers of stomach, liver, colon, breast and endometrium; and sulphur compounds inhibit cancer cells.	8-10, 14-15
4.	*Brassica campestris* Linn. (Sarson, mustard)	Seed oil	Dithiolthiones and isothiocyanates.	Given in tumours/cancers.	8,14-15
5.	*Brassica oleracea* var. *botrytis* Linn. (Phoolgobhi, cauliflower)	Sprout (flowers)	Ascorbigen, vitamins (A, B, C), sulphoraphane and isothiocyanate.	Given in cancers of bladder, lung, stomach, colon, rectum and breast.	7-10, 14-15
6.	*Brassica oleracea* var. *capitata* Linn. (Pattagobhi, cabbage)	Leaf	-do-	-do-	8-10, 14-15
7.	*Brassica rapa* Linn. (Shaljum, turnip)	Root, leaf, seed	Ascorbigen, vitamins (A, C) and Ca.	Given in various tumours/carcinomas.	10,14-16
8.	*Citrus limon* Linn. (Nibu, lemon)	Fruit	Vitamin C, flavonoid, flavone, limonoid, limonene (terpenoid), nobiletin and tangeretin.	Flavonoid, tangeretin, nobiletin, limonoid and limonene inhibit cancer growth and nibu inhibits human breast cancer, metastasis and leukaemia.	8-9

Contd...

Table 1–*Contd...*

Sl.No.	*Botanical (with Hindi or English) Name*	*Part Used*	*Main Active Components*	*Specific Antioxidant/Anticancer Activity*	*References*
9.	*Coriandrum sativum* Linn. (Dhaniya, coriander)	Leaf, fruit	Essential oil, vitamin C, carotene, borneol, limonene and α-pinene.	Antioxidant.	10-11,15
10.	*Cucumis sativus* Linn. (Khira, cucumber)	Fruit, seed	Vitamin C, 14 α-methyl-phytosterol and amyrins.	Given in tumours.	10,15
11.	*Curcuma longa* Linn. (Haldi, turmeric)	Rhizome (Tuber)	Curcumin, β-pinene, camphene, eugenol, curcuminoids and β sitosterol.	Active against esophagus, colon, liver, bladder and prostate cancers and given in leukaemia, fibrosarcoma and stomach papilloma.	8,10-11, 14-17
12.	*Daucus carota* Linn. (Gajar, carrot)	Root, leaf	Carotene, flavonoid, carotenoid and glycoside.	Given in tumours.	10,15
13.	*Glycine javanica* Linn. (Soybean)	Seed	Phytates, saponins, phytosterols, lignans, isoflavonoids and isoflavones.	Lowers the risk of breast, prostate, colon, stomach, rectum and lung cancers and isoflavonoids and isoflavones inhibit many cancers.	8-9
14.	*Lycopersicon esculentum* Mill. (Tamatar, tomato)	Fruit	Vitamins (A, B, C), essential amino acids and lycopene.	Antioxidant and anticancer.	9,14-15
15.	*Mentha spicata* Linn. (*M. arvensis*) (Pudina, garden mint)	Whole plant, leaf	Essential oils (menthol, menthone, limonene), flavonoid and sesquiterpene.	Antioxidant and anticancer.	10,14-15
16.	*Momordica charantia* Linn. (Karela, bitter gourd)	Leaf, fruit, seed	Linoleic acid, palmitic acid, momordin and vitamins.	Active against colon, breast, bladder and prostate cancers and given in lymphoma and leukaemia.	10-11, 14-15,17

Contd...

Table 1–*Contd...*

Sl.No.	*Botanical (with Hindi or English) Name*	*Part Used*	*Main Active Components*	*Specific Antioxidant/Anticancer Activity*	*References*
17.	*Moringa oleifera* Lam. (Munga, Mungana)	Root, leaf	Vitamins (A, C).	Anticancer.	12,15
18.	*Nelumbo nucifera* Gaertn. (Kamal, lotus)	Whole plant	Alkaloids, vitamins and quercetin flavonoid.	Antioxidant.	10,15
19.	*Phaseolus vulgaris* (French bean)	Fruit	Amino acids, anthocyanin and quercetin.	Antioxidant and anticancer.	8,10,15
20.	*Trigonella foenum-graecum* Linn. (Methi, fenugreek)	Leaf, seed	Choline, trigonelline, saponin, amino acids, vitamins and quercetin.	Given in various cancers.	10-11, 14-15
21.	*Zingiber officinale* Linn. (Adrak, ginger)	Rhizome	Camphene, gingerol, zingiberene, borneol, cineol, curcumins and proteins	Antioxidant and anticancer and given in tumours.	8,10-11, 14-15,17

carvone). They increase the tumour latency and decrease the tumour multiplicity[7,9].

There are many vegetable pigments such as flavonoids, carotenoids and anthocyanins which protect us from various diseases. Flavonoids extend the activity of vitamin C, act as antioxidant and have antitumour activities. Quercetin (a flavonoid) chiefly present in red and yellow piyaz and French bean possesses both anticarcinogenic activity. The carotenoids are powerful antioxidants that provide protection against oxidative damage, and stimulate immune function. Persons with high levels of serum carotenoids have a reduced risk of cancer. In addition, a variety of phenolic compounds (*e.g.*, caffeic, ellagic and ferulic acids, sesamol, and vanillin) are present in vegetables. These exhibit antioxidant and anticancer activities. Adrak contains phenolic compounds (gingerol) which have antioxidant activity that is even greater than α-tocopherol (vitamin E). Compounds that stimulate the activity of GST are considered as inhibitors of cancer. Substances that stimulate GST activity are phthalides in celery seeds; sulphides in piyaz and lasun; dithiolthiones and isothiocyanates in sarson, rai, phoolgobhi and pattagobhi; limonoids in nibu; and curcumins in adrak and haldi[1,8–9]. Vegetables and fruits also contain a variety of isoprenoid compounds that exhibit antitumour/anticancer activities. These compounds include tocotrienols (related to tocopherols) and terpenoids (*e.g.*, limonene, geranoil, menthol and carvone). Overall, these compounds increase tumour latency and decrease tumour multiplicity[7,9].

References

1. Madhuri S. Studies on oestrogen induced uterine and ovarian carcinogenesis and effect of ProImmu in rat. PhD thesis. Jabalpur, MP: RDVV; 2008.

2. Pandey Govind, Madhuri S. Global scenario of radiation in cancer risk: A review. Int J Pharmaceu Biol Arch 2010; 1(3):195-8.

3. Madhuri S, Pandey, Govind. Ethnomedicinal plants for prevention and treatment of tumours. Int J Green Pharm 2009; 3(1):2-5.

4. Pandey Govind, Madhuri S. Some anticancer agents from plant origin. Pl Arch 2008; 8(2):527-32.

5. Rosangkima G, Prasad SB. Antitumour activity of some plants from Meghalaya and Mizoram against murine ascites Dolton's lymphoma. Ind. J Exp Biol 2004; 42:981-8.
6. Sivalokanathan S, Ilayaraja M, Balasubramanium MP. Efficacy of *Terminalia arjuna* (Roxb.) on N-nitrosodiethylamine induced hepatocellular carcinoma in rats. Indian J Exp Biol 2005; 43:264-7.
7. Vecchia CL, Tawani A. Fruits, vegetables and human cancer. Eur J Cancer 1998; 7:3-8.
8. Craig WJ. Phytochemicals: Guardians of our health website. 2006.
9. Heber D. Vegetables, fruits and phytoestrogens in the prevention of diseases. J Postgrad Med 2004; 50:145-9.
10. Kaur C, Kapoor HC. Antioxidants activity and total phenolic content of some Asian vegetables. International J Food Sci Tech 2002; 37:153-61.
11. Rao KVK, Schwartz SA, Nair HK, Aalinkeel R, Mahajan S, Chawda R, Nair MPN. Plant derived product as a source of cellular growth inhibitory phytochemical on PC-3M, DU-145 and LNCaP prostate cancer cell lines. Current Science 2004; 87(11):1585-8.
12. Pandey Govind, Madhuri S. Some medicinal plants as natural anticancer agents. PHCOG REV 2009; 3(6):259-63.
13. Ray G, Hussan SA. Oxidant, antioxidant and carcinogenesis. Indian J Exp Biol 2002; 40:1213-32.
14. CSIR. The Useful Plants of India. New Delhi: Council of Scientific and Industrial Research; 1986.
15. Prajapati ND, Purohit SS, Sharma AK, Kumar T. A Hand Book of Medicinal Plants. India: Agrobios; 2003.
16. Jadeja BA, Odedra NK, Solanki KM, Baraiya NM. Indigenous animal healthcare practices in district Porbandar, Gujarat. Indian J Trad Knowledge 2006; 5(2):253-8.
17. Aruna K, Sivaramkrishanan VM. Plant products as protective agents against cancer. Indian J Exp Biol 1990; 28:1008-11.

Index

www.ingramcontent.com/pod-product-compliance
Ingram Content Group UK Ltd.
Pitfield, Milton Keynes, MK11 3LW, UK
UKHW022211051026
15624UKWH00001B/118

9 789351 301653